Why Five Toes Instead of Four?

Did Darwin Make a Mistake?

A Solution to Human Bipedalism

Why Five Toes Instead of Four?

Did Darwin Make a Mistake?

A Solution to Human Bipedalism

Jordi Fuentes

ISBN 1-58721-486-5

1stBooks – rev.03/03/2000

Summary: Connects biology and anthropology. Proposes a new hypothesis about human evolution. Establishes new insights about the origin of our species, and answers anthropological questions not solved until today.

Proofreading: Susan Meigs

Contents

Note To The Reader

In order to avoid repetition, all references taken from Darwin's work are quotations from the book *The Origin of Species and The Descent of Man* by Charles Darwin from the Modern Library, New York, published by Random House, Inc.

In regard to each individual quotation, only the number of the corresponding page is given.

Introduction

An introduction? Oh no! Not an introduction. Please! Nobody likes introductions. Who reads them? Curios, isn't it?

My research on this has produced an unexpected result. Only one person in six reads the introduction of the book he or she selected to read. Why? I have no answer to this question. But believe me, many, many people skip reading the introduction. I have asked a lot of people the reason why, but I have received only nebulous answers, the most common being that "they are interested in the contents of the book, and not in what the authors talks about before entering into the subject matter." Is that a reasonable answer? Perhaps for someone, but to me, it appears that people who do not read the introduction do not understand the meaning of "introducing" someone to an unknown subject. Well, let's stop wondering about the intricacy of the human brain, and let's go directly to the point.

My intention here is to show the reader the purpose of this book and my intention in writing it. The objective of this work is to help us understand ourselves a little bit more. The only way that my limited mental capacity allows me to achieve this objective is to study the present knowledge that we have about ourselves and adjust the concepts. The evaluation of my exposition is something that each one of the readers has to ponder by himself. My participation is only to present the results of my work, and then let everyone arrive at his or her own conclusion. I cannot, and I do not want to, impose ideas. I only present here the facts that I consider closer to the truth. Besides, no one has a monopoly on truth.

The trouble with truth is that it is not absolute, as it depends on how much we are ready to accept. Our willingness to accept something as a fact depends on our subjective acceptance, which in turn depends on our personal experiences and the instruction received throughout our lives.

Nothing is right or wrong. All is changing continuously, physically, and conceptually. What was reality for people who lived two thousand years ago is not the same reality we accept today. What had an enormous value for them, today is completely meaningless.

Concepts change, ideas change, and experiences change continuously according to the changing moments, periods, or epochs. Nothing is stable. When we think that we have a cosmic reality, pretty soon we realize that this reality changes and is only valid for us and for an insignificant period of time. One's life? One's life is over in no time. Eternity? What's that? Concepts change and are valid only for a moment; it doesn't matter that the moment is measured in centuries or millennia. Anyway, a moment is only a moment; it doesn't matter how long we think it is. It is a concept, and concepts are variables without reality. Their presence is in our mental computer without a physical representation. Once the mental computer is disconnected, the concepts stored in it disappear completely. Is a physical singularity a reality? Believe it or not, physical realities change from the moment they are produced. An iron bar changes? Yes it does. From the very moment that we produced the iron bar, chemical reactions occur in its outer layer. Those changes could be very slow and apparently imperceptible, or, rather, imperceptible to our senses, but they are there. Ideas change also. What is logical at this moment could be absurd sometime later, when new experiences and different facts produce another idea or ideas that invalidate the first one.

We are here, we affirm. Is it an affirmation? Are you sure? What does it mean to be here, in this spot? Well, this spot actually does not exist. One can pinpoint the spot on the ground in relation to other physical points, but while we consider this position in space, the Earth has moved several thousand miles; the spot that we consider, even if it maintains the same relationship with the other points, has moved the same distance that the Earth did, in relation to the position in the solar system. The solar system is moving at an enormous speed inside our galaxy (twelve miles per second), in the direction of the bright star, Vega. In addition, since our galaxy is rotating, our sun is also orbiting the galaxy's center. Our galaxy, the Milky Way, which is part of a cluster of thirty galaxies, is moving vertiginously (fifty miles per second) toward the Great Andromeda galaxy. This cluster of galaxies is speeding in the direction of the Virgo cluster of galaxies at 150 miles per second. And if that is not enough, everything is expanding in the universe at some 322 miles per second. Where is our initial spot? Nowhere. Everything changes. We change. Our ideas change. We think that because today we have an idea, this idea is going to be valid 100 years later? What is 100 years? In relation to human life, 100 years could mean a long time; but, is it? In terms of the cosmological scale, a 100 years is absolutely nothing. The value that we give to ideas and to material things, last in relation to our lives, and our lives are so short . . .

Darwin was a pioneer in the study of man considered as a creation of nature. My opinion of his work: If I had to evaluate it from one to ten, I could give him: 100. Yes, I consider his work a fabulous breakthrough in the understanding of our position in the biological concert.

My work here tries to be a polishing tool to smooth concepts and open new sources of understanding concerning the points that were not completely clarified in Darwin's work. At first sight, the reader might get the impression that my purpose is only to criticize him and his conclusions, which is far from my true intention. The reader has to understand that in order to correct a defect, or polish an idea, the first condition is that the defect or idea has to exist; the second condition is that the defect has to be pinpointed, otherwise, it is impossible to carry out the improvement or polishing. In this case, I am forced to pinpoint the situations in which some improvement can be made in Darwin's writings.

Many people do not want to accept evolution, as we will see in the course of this work, and the reaction against its acceptance is just a misinterpretation of Darwin's work. This misinterpretation was then fortified by religious ideas accepted and repeated hundreds and hundreds of years without reflecting upon them.

To deny the existence of evolution is to deny one of the most ingenious, marvelous and beautiful creations of God.

Before continuing with further discussion, I have to state that I do not agree with some of the vocabulary commonly used in Darwin's work, that was widely accepted in his time. For instance, he uses the word "savages" in reference to less technically developed cultures. According to the definition given in *Webster Dictionary*, "savage" means "*wild; untamed; fierce; cruel; pitiless; inhuman; brutal; without civilization; primitive, barbarous; ferocious; merciless.*"

Less technically developed cultures do not have to be inevitably ferocious, cruel, and brutal. We do not consider ourselves primitive, but is it not brutal to drop tons of bombs over a city and destroy thousands of human beings? But we do not consider ourselves "savages." Human logic. On the other hand, there are cultures with little developed technical advances, that haven't even developed the word for war. They are gentle people, and their violence is only displayed in hunting, necessary for them to eat. Among human groups that we consider primitive because they have a less developed material life, in reality their processes of thought are often more complicated than ours. However, because they are closer to the state from which all mankind once emerged, we name them "savages." I don't agree with the use of this adjective. I never use the word "savage," and when I refer to "primitives," I call them "less developed materialistic societies" or "cultures less developed technologically."

Another obscure point in the vocabulary used by Darwin is that he does not explain what he considers "lower animals" and "higher animals." He uses this classification very often, but has not provided a definition for those terms.

In order to finish this exposition to introduce my work, I need to mention the reality of the state of mind prevailing in the scientific community, which is characterized by intellectual arrogance. However sometimes it takes amateurs or outsiders to see a complex situation clearly. Generally, some individuals' intellectual arrogance and ego glorification do not allow for these individuals to give an unbiased investigation; instead, they completely oppose any other school of thought. Dr. Linus Pauling, twice the winner of the Nobel Prize, said "Science is the search for truth. It is not a game in which one tries to beat the opponent to do harm to others." Unfortunately, the reality is that professionals often are conditioned by prevailing perspectives and technical considerations that tend to block a holistic view and fresh ideas, preventing them from accepting even the most logical explanation. Professional jealousy, plays an important role in the frequent rejection of new and unexpected ideas, mostly when they come from someone not considered a sacred cow in their field. The outsider has profaned the most sacred doctrines jealousy guarded by the high priests or monopolists of the knowledge.

The high priests or guardians of the arcane truth, the "scientists" have, in terms of the following subject matter, complicated things when they considered the origin of man as a split branch of the primates once these animals had already conquered life in trees. It seems to me that the origin of man preceded this point, coming from a very different starting point, as we will see throughout this book. Am I right or wrong? That is the question.

The reader has to be the judge in this case. If, after reading this book, the reader considers that my conclusions make sense, this will be my reward for many years of research. If, on the contrary, the judgement is negative, I have no other recourse than to beg your pardon and continue for the rest of my life with a paper bag on my head.

The author

I

Some Words before Entering the Subject of This Work

Human beings have a tendency to mold the world around them according to their thoughts, instead of adapting themselves to the world. Human beings like to play God. They think that they can change the rules established by nature, and try by all means to create a human world to substitute for the natural world.

The arrogance of man makes him think that he is able to create a better world and dominate nature. He thinks that nature is not perfect, that he can perfect nature. What is the result of this human action? So far, the only thing accomplished is the near destruction of the forests, pollution of the air, water, and land; destruction of the ozone layer, the depletion of the oceans; the calcification of the land with the use of cement to obliterate the surface of the Earth; and the pollution of the atmosphere with all the trash left behind by obsolete satellites, rockets, and likewise. Is this the human species' marvelous accomplishment? And, as if that were not enough, man, driven by his never satiated curiosity, is trying now to pollute other planets of our solar system!

Man is paradoxical. We have not yet learned how to take care of this planet; nevertheless, we are trying to control the other ones in our solar system. Humans have many defects, but playing God is the worst. I am also a human being, so probably suffer from the same ills even if I don't realize it (nobody does). Self analysis is impossible, and self correction even more distant.

The arrogance of man has no limits. We produce an idea and we try by all means to find the way to justify it, whether it agrees with reality and logic or not. If it does not agree with reality, we ignore the evidence and, through an intricate and, most of the time, illogical and elaborate exercise of thought, convince ourselves that we are in possession of the truth. This is self brainwashing. We convince ourselves to the point that anyone who does not accept our idea 100 percent or who does not share our point of view is considered absolutely wrong; and most of the time we classify him or her as stupid, ignorant, or primitive. Any doubt about our position means that we are subject to judgment. We don't like to be judged, we want to be openly accepted. This is human intransigence at work. This is human nature. Anything except tolerance.

When did man begin to consider himself the ruler of the world? We don't know. The only thing that seems, so far, most probable is that the most ancient human thoughts originated and were directed to answer a bunch of questions that man's ascension to intellectuality posed. These questions completely invaded his brain and became his constant preoccupation and worry, in his search for a convenient explanation that could satisfy his ego. Man always tries to demonstrate that his desires will be satisfied in complete accordance with the reality of the world around him, even if the reality is completely different from his thoughts or opposed to his ambitions.

Which questions did the most remote ancestors of man begin to ask themselves, from the very moment that their brains evolved enough to set forth doubts and questions? What are the main ideas that torment most human beings? Practically, it is only one; it is shared by almost everyone in all the different races, cultural groups, and social strata, and, it is also found in all levels of intellectual development. No other idea has permeated the thought of the human being more than **the search for an answer about his future after death**.

Since the very moment that the human being awoke to consciousness, he realized that he was in possession of an existence that was, unfortunately, limited by death. He did not like this reality; he could not accept this fact of life and rebelled against it. He felt the imperious desire to perpetuate his life. He tried to find an answer that could obviously accommodate his desire to continue his life after physical destruction, and forever. If this extended existence could not be in the earthly dimension that death was clearly denying him and all other existing creatures, the human being was at least looking for the continuity of his life in another dimension. This new state, according to his wishes, must be like the present one. . . . Or why not another state of life even better than the physical existence, where suffering, pain and uncertainty permeate our lives from the very beginning? If we have to find a solution to this problem why not search for an answer that could produce an even better life than the material state that we are forced to live in because of our appearance in this physical world?

If we have to leave this world, why not ascend to another one even better? Man is an innate merchant, and he is looking always for bargains.

Before continuing on this subject, let me make an introduction for our better understanding. Human beings are polarized, or, rather, let's say bipolarized. All our existence and thoughts are governed by two dimensions, or poles: up and down, father and mother, dark and light, right and left, good and bad, young and old, beautiful and ugly, dry and wet, tall and short, large and small, happy and sad, friend and enemy, and so on. The list is enormous. Two poles are the idea chiseled in our brains. Everything has to have a counterpart (or at least we think so). We evolve between two positions or dimensions. Yes, there are other dimensions, but for most of us, including myself, it is very difficult to think in a third dimension, even if we clearly accept its existence.

According to this bipolarization of our brain, everything has to have a beginning and an end. The human mind can accept this. Nobody questions that everything is born, grows, changes, and disappears. Everyone accepts continuous change from the beginning until the end. Mountains are formed, grow, erode, and disappear. Rivers and lakes are also formed, have a moment of glory, and, through the centuries or millennia, disappear, gone forever.

In reference to the undisputed continuous changes of the surface of the Earth, Darwin comments (page 372), *"The noble science of geology loses glory from the extreme imperfection of the record. The crust of the Earth with its embedded remains must not be looked at as a well-filled museum, but as a poor collection made at hazard and at rare intervals."*

Everyone accepts the continuous change of the Earth, and the creation and destruction of all the geologic formations, and with them the fossil records of all the generations of plants and animals that have lived on this planet for millions of years. This vision confirms that nothing in this world is stable. Everything changes. Not even the periodic movement of the Earth is regular. The Earth cannot return to the same point, because while the Earth rotates around the sun, all our solar system is changing position in our galaxy, the Milky Way. At the same time, the Milky Way is moving impetuously toward another galaxy at an incredible speed. That means that the point in space that Earth occupies at this moment will never be reached again. Everything changes. Nothing is stable. Everything is in continuous movement and transformation.

Animals and plants are, obviously, born; they grow and die. Everything acts according to this bipolarization: a beginning and and inexorable end. But . . . . beware! Our life to have an end? Death! Oh No! That is not in accordance with our desires! We had a beginning in our existence, that is OK. We accept that. But by no means should we have to have an end! That cannot be right! We have to continue existing forever! That is what we want. How can we disappear from this scenario in which we are moving, feeling, and enjoying, and in which we want to perpetuate forever? No way! There has to be another solution! Human haughtiness in action.

We accept that everything has a beginning and an end, but, oh no we have to be an exception! It is OK that everything has to finish somehow, someday, but, no we cannot finish! We easily accept that we had a beginning, but an end? No way! We have to be an exception to the bipolarism imprinted in our brains.

Here, the human being began to work overtime to find an escape from his bipolarized mind. But what escape?

From the very beginning of the human ascension to intellectuality, the ingredients were ready for the mental concoction that our brain was preparing in order to conform ourselves, justify our existence, and satisfy the desire to perpetuate ourselves forever.

Everyone in every group, in every culture, in every country everywhere has the same desire to perpetuate his existence. Why not? Please do not confuse the desire to perpetuate oneself with the conservation instinct. The first is a product of our mind, the second is a reaction to the circumstances, inherited in the genes of any living creature, and helped, in our case, by our suprarenal glands.

Everyone is definitively interested in finding a way to continue living; everyone is interested in the future after our material disappearance. What will happen after our death? All our efforts, thoughts, and behavior are focused to assure ourselves that our existence will continue after death forever. We cannot accept another solution. No way! Why should we have to? Human logic at work.

In the search for this anxiously needed answer to this imperious question, human thought, at last, solved the problem. The human mind began to formulate another dimension where we could continue the existence that obviously was denied to us in this material world. But how can we reach this end? If we cannot do it ourselves, because we are conscious that in the beginning Someone had created us, then the most simple solution is to resort to this Someone and, with his help, resolve this crucial situation! Simple!

Obviously, the one who is so powerful as to be able to create this world and ourselves (remember that we had to have a beginning) is the one with the power to produce the solution we are looking for. In our mind there was no doubt that everything had to have a Creator or Maker. We cannot think otherwise. Here, then, is the grand solution: A Supreme Being, the same one who granted us this life, can grant us the continuity of it after death. That way, man found a solution to his worries: God.

But how to get in touch with such a Supreme Being? There has to be a way, a means of communication through which we can keep in contact with the Ruler of Creation and tell him about our concerns and our desires. Then . . .

The human mind created religion.

Yes, religion is a product of the human brain. Where there are no humans, there is no religion. Where there are no humans, there are no feelings, thoughts, desires, or the need for eternal existence. Where there is not a mental computer, our brain, no explanations are needed.

But what is religion? Religion is the most difficult thing to define. There have been thousands of definitions, but not one of them includes all the ingredients found in all the existing religions, to the point that any proposed definition of religion leaves out those religions that lack some ingredients or have other ingredients not mentioned in the definition. Perhaps the best definition of religion would be "man's attempt to achieve a state of supreme goodness or perfection, reconciling his life to the strongest and best power in the universe." This power has generally been called God.

Human beings found a haven in religion, a place to take refuge against the repudiated idea of their disappearance from an earthly existence for them, impossible to ignore. The physical evidence of the destruction of our bodies forced the human being to find assurance that his destruction is not permanent, only a passage from one form of life to another. The destruction of the body, we like to think, is only a facet or step toward an eternal existence, which, in many instances and helped by our desire and by an excessive imagination, is even better than the present life. Almost all religions possess an imaginary heaven or paradise.

If the body is destroyed by death, then reality forces us to accept this. How can man perpetuate his life after death? When facing this situation, the human mind was forced to produced a solution. He developed the idea of something not material hence, not subdued by death's material manipulation and which, by its own immaterial nature, could survive the final stage of the body on Earth.

The idea of a soul was born.

It is very curious that the human being is so worried and preoccupied about where he will be after death, but nobody worries or is preoccupied about:

Where were we before we were born? Did we have a soul before being born?

No one is concerned about our past. It is easily accepted that we did not exist in the past, but the assumption that we could not exist after this period of consciousness of our existence is an intolerable and inadmissible idea. Strange isn't it? The past doesn't count. It doesn't affect our feelings, it cannot affect our future. It's only history. But voilà! Everyone is worried about the future. . . .

The idea of eternal existence has to be consistent with our present life. That is to say, the life we want to perpetuate has to be, at least, the same good life we enjoy now, or perhaps even better. Why not? But by all means not worse. Commercialistic human nature does not allow us to settle for something of less or the same value, if we have the opportunity to get something more or better for the same price.

If we want to continue living in a body like the one we are living in now, nature has to be static and by no means evolving from one being to another. If we evolved from an inferior form of life, it means that in the life after death there would be all kinds of souls from the souls of the most primitive creatures to the souls of developed people like us. In this life after death, we have to move through all the stages of development of souls from the souls of inferior creatures such as a paramecium to the souls of modern man. Does the Cro-Magnon man have a soul? and the Neanderthal? And the *Australopithecus*? And all the other hominids? Until when can we go back in time finding souls in our ancestors? Since this idea appears impossible to accept, then it is better to forget about it and close our brains to such absurd reasoning. Why worry about this thought, when it is easier to forget about the past and look only forward? So the idea of evolution is in opposition to many of the more than 1,700 religions in existence today.

Some religions accept evolution as a stage in a long journey evolving from a certain level of life to a superior one. The good behavior and rightness during one's life direct the individual to a better one. Perfection is obtained through different stages of improved behavior. The physical body can change in each stage, but the individual follows his march toward an ideal ending of perfection and sublime happiness.

Idyllic, isn't it? Yes, human beings have a very active and prolific imagination, which, under the effects of repetition, becomes a reality in our minds. The imprinting of religious thoughts and ideas produces hope in the individual, and because of it they are easily nailed down in our brains. No other idea is more eagerly accepted than the promise of a happy and eternal existence. That is what man is looking for: hope. Hope is the product that all religions are offering, hence, their success.

The lack of proof regarding the existence of such ideas doesn't prevent or influence their acceptance. The inner desire to find a solution to our worries opens the way to acceptance without hesitation or questioning its possibilities. If someone dares to question the possibility of the existence of such possibility, he soon succumbs to the thought "just in case." And if someone persists in questioning, he is considered our enemy, a pawn of the devil.

The offered hope is enough gratification for man, who not only moves away from the possibility of doubt, but energetically rejects any question or idea that can cloud the offered solution to his most acute mental problem. Because of the profound feelings that religion causes, different religious points of view have brought intolerance between different faiths. This is the origin of religious intolerance, which paves the way for all manifestations of religious fanaticism and, at the same time, gives a good reason to practice for a "good cause" our innate aggressiveness (see the chapter "Human Aggression".)

Darwin remarks about religion (page 468): *"There is no evidence that man was aboriginally endowed with the ennobling belief in the existence of an Omnipotent God. On the contrary there is ample evidence, derived not from hasty travelers, but from men who have long resided with savages, that numerous races have existed, and still exist, who have no idea of one or more gods, and who have no words in their languages to express such an idea.*1 *The question is, of course, wholly distinct from that higher one, whether there exists a Creator and Ruler of the universe"* (page 469). *"The belief in spiritual agencies would easily pass into the belief in the existence of one or more gods. For savages would naturally attribute to spirits the same passions, the same love of vengeance or simplest form of justice, and the same affections which they themselves feel."*

Personally, I do not agree with Darwin's point of view, because great civilizations such as the Greek and Roman civilizations cannot be classified as "savages," and they believed in many gods, which had the same passions and feelings as the human beings. These Greek and Roman gods had affections, love, and the feeling of vengeance to the point that they married to have descendants, and fought against each to solve personal disputes.

It is dangerous to generalize about something so intimate and widespread as religious feelings that, in many instances, mix fantasy with reality. In very well developed western cultures such as those found in European countries, some forms of religion accept the ascension of human beings to sanctity, a midway point between the common man and the Supreme Being: God. In those cases, people believe that these elected human beings, and even their images, can act in their favor and intercede for them to get favors and privileges from God. A sort of lobbying. If that happens in advanced cultures, it is no wonder that less evolved populations have the same or analogous thoughts.

During my anthropological and linguistic studies with the inhabitants of Easter Island in the South Pacific, I found that they had only a remote idea of a Supreme Being, to the point that we can consider it an esoteric idea and not a common concept. This island belongs to the Republic of Chile. During the last quarter of the last century, this island began to be influenced by the Catholic religion through missionaries. Before their teachings, the closest expression to Supreme Being was the word *"manaranga,"* meaning "that which exists by itself; or that which has been created or formed all by itself." That "it" had no physical representation. A later introduction of the concept of God from the Tahitian islands is found in the expression *"Atúa,"* meaning *"someone."* Probably *"Atúa"* is a derivative of the word *"Atu,"* meaning "toward there," and *"A"* meaning "to exist; something or someone that exists there." They hadn't even developed the idea of a bad spirit. The word *"Tatane"* a derivative of the word *"Satan,"* is a recent introduction to the island, when Christian missionaries began to teach Christianity to the island natives.[2]

Darwin says (page 470): *"The Fuegians appear to be in this respect in an intermediate condition* [he refers to the belief in spiritual agencies], *for when the surgeon on board of the "Beagle" shot some young ducklings as specimens, York Minster declared in the most solemn manner, 'Oh, Mr. Bynoe, much rain, much snow, blow much;' and this was evidently a retributive punishment for wasting human food, so again he related how, when his brother killed a 'wild man,' storms long raged, much rain and snow fell. Yet we could never discover that the Fuegians believed in what we should call a God, or practiced any religious rites; and Jemmy Button, with justifiable pride, stoutly maintained that there was not devil in his land. This latter assertion is the most remarkable, as with savages the belief in bad spirits is far more common than in good ones. The feeling of religious devotion is a highly complex one, consisting of love, complete submission to an exalted and mysterious superior, a strong sense of dependence,* [3] *fear, reverence, gratitude, hope for the future, and perhaps other elements."*

During my experiences with the last remnants of the Alacaluf Indians in the Aysen region of Chile, I could never find any traces of religion or beliefs in life after death. They accepted this life as the only existence in this world, and they had no spiritual feelings or connections with invisible beings. It was very difficult for them to interpret the teachings of the Christian missionaries that tried to convert them. Their world was the world around them, without any connection with a spiritual or invisible dimension. They accepted tangible things as the only reality. They did not fear evil spirits, because the idea of something immaterial had not been developed in their culture. They accepted nature as the only existing reality, and if man made some offense to nature, nature alone, no spirits or deities, could revenge. Their rituals for a deceased person involved only an elevated platform of logs or branches about two meters from the ground, and they left the body there for nature to recycle it.

Not all individuals share the same degree of religious sentiment. Some persons are very active participants in the practice of the religion imposed by their culture. Some others participate only occasionally, more as an established routine or custom that transforms the religion into a habit than as a spiritual need. Other people act because they are driven by a strong feeling of superstition. It is very difficult to draw a line between superstition and religion. In many instances, they overlap, and in this case, it is not possible to establish where one starts and the other finishes.

Some religion accept rituals and practices involving magic as part of their liturgy. In most cases, this is the product of mixing two different religions, as happens in places where the Christian religion spreads to areas of native religions or where foreign religions are established through the emigration or import of people of a different culture. This example is often observed in countries of the Caribbean, and Central and South America.

The religious sentiment or feeling is not equally shared among individuals of different life stages. During childhood, the individual does not have a clear idea of what religion means. Children just follow the example of the parents, without a deep knowledge of what they do, or the meaning involved in it.

During adolescence, the individual usually suffers a difficult transition from infancy to adulthood. During this period, leaving apart some exceptions, the person tries to act as adult, but does not have the mental maturity to act as such. Very well known is the fact that most teenagers suffer accidents that could be avoided if they had a more mature reasoning capacity. This stage of development creates in the adolescent a false feeling of security. He thinks that he has reached a clear understanding of the reality around him, and tries to act as an adult without the backup of adult maturity. In this period, religious sentiment is almost null. A teenager thinks that he knows everything and that he is the owner of the world, if not the universe. Remember the saying: "Why didn't the problems of life hit me when I was sixteen and I knew everything?"

At the same time as the person ages, he or she unconsciously has the notion that he or she will vanish from this life, and the religious feeling begins to take shape and increase in the individual's mind. As the person grows older, material needs dwindle rapidly and food for the mind becomes more imperious than food for the body. As a consequence, the person becomes more and more mystic. If a mental disease, so often developing in aging people, does not interfere, an older person's last years are the ones in which mysticism completely permeates his life, and he dies with the feeling that complete communion with God has been achieved. All earthly faults have been pardoned and the way to an afterlife is guaranteed.

If we want to understand the reaction produced by Darwin's teachings, which in his time, were opposed to the generally accepted reality formulated by the religious teachings of the Jewish-Christian culture, we have to consider the profound impact that these old concepts had produced throughout thousands of years.

It is not easy to change all concepts that, in that case, played such an important role in the already established ways of life. We have to be flexible in accepting the criticism that Darwin caused which originated due to lack of reflection, and to widely spread concepts accepted by the multitudes that never had access to Darwin's works and who condemned them without even reading them.

Most people affirm that Darwin said that man descended from the monkeys. Nothing is further from the truth. What Darwin said is (page 444): *"In regard to bodily size or strength, we do not know whether man is descended from some small species, like the chimpanzee, or from one as powerful as the gorilla; and therefore, we cannot say whether man has become larger and stronger, or smaller and weaker, than his ancestors."*

Prejudice, the act of judging without knowing the facts, is, I should say, part of human nature. It is easier to reject the facts than to make the effort to see if in reality, the facts deserve rejection or if they possess some value in themselves.

Notes

1. Darwin refers readers to an excellent article on this subject by the Reverend. F. W. Farrar, in the *Anthropological Review*, August 1864, page ccxvii. For further facts, see Sir J. Lubbock, *Prehistoric Times*, 2nd Edit. 1869, page 564; and especially the chapters on religion in his *Origin of Civilization.* 1870.
2. *Dictionary and Grammar of the Easter Island Language,* Jordi Fuentes, Santiago de Chile, 1960.
3. Darwin refers the reader to an article on the *Physical Elements of Religion*, by Mr. L. Owen Pike, in Anthropological Review, April 1870, page LXiii.

II

Darwin's Proposed Theory

When we use the word "evolution" we mean "organic" evolution, or the "theory of evolution" applied to living things. This theory, first published by Charles Darwin, involves the idea of biological evolution. The definition of *biological evolution*, according to *Webster's New Universal Unabridged Dictionary*, is: *"The development of a species, organism, or organ from its original or rudimentary state to its present or complete state."*. This process includes *ontogeny* and *phylogeny* of all organisms involved. First, let us note that the definition says: "from its original state." Do we know which was the original state? No. Absolutely not. What does "the present or complete state mean"? Does that mean that the present state is the final stage of evolution? I sincerely think that is not so. Some scientists have said that evolution has come to a standstill in the present, suggesting that what has taken place for millions of years has ceased to act in the present. I do not agree with this idea. First, in a human life there is not time to notice any changes caused by evolution in organisms that took millions of years to produce. Evolution is slow compared with the human life span. Second, evolution is presently working with all its power, and a good example is that bacteria that cause some diseases change their appearance continuously, becoming resistant to the different antibiotics used against them. They change from one season to another. What about the cold virus, flu virus, and so on? Here the same situation is repeated. Viruses change, continuously adapting themselves to new situations produced by our interest in suppressing them. It is a continuous fight. It is evolution at work. It is a fight in which we cannot see, so far, the way to win the battle.

The theory of organic evolution involves three major ideas. First, living things change from generation to generation, producing descendants with new characteristics. Second, this process has been going on so long that it has produced all the groups and kinds of things now living, as well as others that lived long ago and have died out, or have become extinct. We know them by their presence as fossils. Third, these different living things are related to each other, and so are the families or larger groups of which these various living animals or plants are members.

With respect to his theory, Darwin says (page 66): *"It may metaphorically be said that natural selection is daily and hourly scrutinizing, throughout the world, the slightest variations; rejecting those that are bad, preserving and adding up all that are good; silently and insensibly working, whenever and wherever opportunity offers, at the improvement of each organic being in relation to its organic and inorganic conditions of life. We see nothing of these slow changes in progress, until the hand of time has marked the lapse of ages, and then so imperfect is our view into long-past geological ages, that we see only that the forms of life are now different from what they formerly were."*

If we remember what I said in the previous chapter, we see that the human mind is so permeated by the ideas established during millennia of efforts in search for eternal life, that it is not surprising that a reaction is produced against any new idea that is not a 100 percent in accordance with the established dogma. Radical changes are not accepted easily. The mild opposition to Darwin's ideas to his first edition of *The Origin of the Species* increased enormously and bitterly with his other book *The Descent of Man*. This outrageous reaction forced Darwin to adopt a more reconciliatory position, but standing his ground, and in his own words, he says (page 441 and 442): *"I have altered the fifth edition of The Origin so as to confine my remarks to adaptive changes of structure; but I am convinced, from the light gained during even the last few years, that very many structures which now appear to us useless, will thereafter be proved to be useful, and will therefore come within the range of natural selection. Nevertheless, I did not formerly consider sufficiently the existence of structures, which, as far as we can at present judge, are neither beneficial nor injurious; and this I believe to be one of the greatest oversights as yet detected in my work. I may be permitted to say, as some excuse, that I had two distinct objects in view; firstly, to shew that species had not been separately created, and secondly, that natural selection had been the chief agent of change, though largely aided by the inherited effects of habit, and slightly by the direct action of the surrounding conditions. I was not, however, able to annul the influence of my former belief, then almost universal, that each species had been purposely created; and this led to my tacit assumption that every detail of structure, excepting rudiments, was of some special, though unrecognized service. Anyone with this assumption in his mind would naturally extend too far the action of natural selection, either during past or present times. Some of those who admit the principle of evolution, but reject natural selection, seem to forget, when criticizing my book, that I had the above two objects in view; hence if I have erred in giving to natural selection great power, which I am very far from admitting, or in having exaggerated its power, which is in itself probable, I have at least, as I hope, done good service in aiding to overthrow the dogma of separate creations."*

Many people did not understand the new idea, and even today there are thousands of people who do not understand it. It is curious that many people reject Darwin's idea of natural selection, but if you ask them if they have read Darwin's work, you will find out that they have not. They reject something that they don't know, and without expending the least effort or curiosity to see what it is all about. It is the same position of people who when offered a certain food, reject it saying, "No thank you; I don't like it." And if asked if they have tried it before, they say emphatically, "No." How can they know that they don't like the offered food, if they have never tasted it?

The evidence for the theory of evolution is based on several kinds of evidence. The following five types are most notorious:

1. **The variation or change**. Everybody knows that all kittens born to the same cat in the same litter are not alike. These variations are the result of the presence of the hereditary message contained in the genes, which combine to produce the differences between siblings. The differences are transmitted from one generation to the next, sometimes without producing any visible change. But the information is still there, and can surface when the conditions are favorable, as when paired with the appropriate information received from another parent. Variations also occur spontaneously as the product of mistakes in the combination of information contained in the genes. These changes or "mutations" can also be perpetuated by heredity. The variations that occur in one individual, as a consequence of an error during the formation of the new being, are called somatic variations, meaning variations in the soma (body), not in the genes, which would be genetic variations. Somatic variations are not hereditary, but they are taken advantage of profusely in agriculture using vegetative reproduction.

2. **Fossils.** Fossils are the remains or traces of creatures that lived ages ago. They are preserved in rock layers called strata, which are the remnants of layers of materials deposited on the dead creature. Fossils are the most important tools that we possess to detect what animals and plants lived throughout the ages. Fossils show the evolution of different living beings from one stage to another, demonstrating the branching series that started from the simple forms, evolving gradually into more and more complex organisms found in later strata. Evidently this record is not complete, but as more fossils are found, new pieces to the evolutionary puzzle are

added continuously.

3. **Embryology**. Much of the information that we have today about evolution, comes from the studies in **embryology**. These studies show the similarities between the embryos of the most diverse animals during the first stages of development, and the different processes of differentiation that happen during the formation of the fetus.

4. **Comparative anatomy.** Comparative anatomy is the other tool that explains the origins of structures present in animals, showing their relation and the changes that had occurred during their evolution until the present arrangement in the animals today. Comparative evolution also explains the origin of structures that now are small or useless.

5. **Geographic Distribution.** Geographic distribution holds much evidence about evolution in species that had become separated for a long time. Darwin found much of the evidence to formulate his theory of evolution in the geographic distribution of animals and plants that live in faraway and different places. The differences that he found in the animals living on islands far from their relatives on continents confirmed his assumption that these creatures were related and that the differences between them were a consequence of their confinement, hence their separate evolution, giving place to notorious structural differences as adaptations to the new environment and sources of food. Geographic distribution supports evidence of the other three main ideas in the theory of evolution. Fossils found in strata in different locations show the variations and changes through which the species evolved from their ancient ancestors. Quoted from Darwin's writing is this sentence (page 67): *"But in all cases natural selection will ensure that they [the modifications] shall not be injurious: for if they were so, the species would become extinct."*

Most people do not realize the difference between evolution and natural selection. Many people think that they are two names for the same thing. Big mistake. The same people who reject the idea of evolution without knowing or reading Darwin's work accept that their son resembles the father, the mother, or some other relative. They accept that the resemblance or characteristics persist in the family through inheritance. Well, that is the way evolution works. These same people might also have a dog. If you ask them what breed the dog is, they, without hesitation, will answer, "Pointer, terrier, poodle," or whatever breed they think it is. Well, with this answer they unconsciously are saying that they accept the principle of evolution, as all the different breeds are the result of the selection and perpetuation of natural variations (called *mutations*) considered desirable for the purpose that the breeder has in mind. That is evolution put to work by the breeder. A quotation from Darwin says: *"We have good reason to believe, as shown in the first chapter, that changes in the conditions of life give a tendency to increase variability."* (page 63).

Natural selection is the preservation of those features that benefit, to some degree, the life of the creature, animal, or plant. Nature is in charge of preserving the individual who has some advantage with the possession of such characteristics, and destroying the individual who has characteristics detrimental to its life. That simple.

How can we explain the many existing and different races of humans if evolution did not exist? Can we accept that every human race was created originally and independently of the others? Or, we have to accept that somehow the original man and woman, Adam and Eve, evolved into many different races? If we accept this last possibility, then we cannot deny the existence of evolution.

Continuing with this line of thought, Darwin said (page 416): *"If we consider all the races of man as forming a single species, his range is enormous; but some separate races, as the Americans and Polynesians, have very wide ranges. It is well-known law that widely-ranging species are much more variable than species with restricted ranges; and the variability of man may with more truth be compared with that of wide-ranging species, than with that of domesticated animals. Not only does variability appear to be induced in man and the lower animals by the same general causes, but in both the same parts of the body are affected in a closely analogous manner. This has been proved in such a full detail by Gordon and Quatrefages, that I need here only refer to their works."* 1

Ancient Greeks knew that the Earth was round. The Polish priest Nicholas Copernicus (1514 A.D.) proposed the idea that the sun was stationary at the center and that the Earth and the planets moved in circular orbits around the sun. Galileo supported Copernicus' idea, and since then the idea has been proved true with some changes, of course, such as the elliptic orbit of the planets instead of the circular one, and the movement of the sun through our galaxy.

As I have said before, radical changes are not accepted easily. Today, after hundreds of years, still there are people who question than the Earth is round, and that the world moves. Everyone has the right to think as he pleases! It is no wonder that an idea such as evolution, of only 150 years old, has not been universally accepted yet, especially when it goes against the religious teachings that, for thousands of years, have said otherwise.

It is not difficult to understand that a new idea such as the one proposed by Charles Darwin in his book about the evolution of the species, entitled *On The Origin of the Species by Natural Selection or The Preservation of Favored Races in the Struggle for Life"* (1859), was not welcomed by the public at large. Evolution was against the religious ideas held for thousands of years and sustained by religious groups, even today. This opposition was fueled by the same Darwin who made the mistake of not properly applying his own theory to the human species. If Darwin had studied the evolution of the human being more precisely in all its aspects and had accurately applied his theory to man before publishing his book, probably the opposition would not have been as negative as it was.

I will explain in this work the points in which Darwin failed in applying his theory to man.

Darwin did not consider two vital characteristics of the human being not found in our cousins the anthropoids. They are the fundamental aspects that differentiate us from the rest of the primates. We will develop these two features as the main reason and purpose for the present work.

1. **The little developed human newborn**
2. **The existence of five toes on each foot instead of four**

These are two aspects of the human being that Darwin did not take into consideration, and his omission has not been noticed by science yet. Darwin followed the idea that we are the descendants of some arboreal primate. He did not realize that our foot was never a hand (see the chapter on Darwin's Second Omission: Man's Foot Was Never a Hand). The present work tries to develop these two characteristics so openly ignored, and which give light to our origin.

What is a little more difficult to understand is why Darwin has been bitterly criticized by many of the scientists who recognize the existence and values of evolution. Their arguments are that Darwin did not solve all the aspects of evolution. We have to consider the fact that Darwin was a pioneer in this field. It is impossible to study such a vast field of science in a man's life; besides, Darwin was not aware of Mendel's work concerning the laws of heredity in plants and applicable to animals as well. Consequently, Darwin did not know the genetic discipline developed as a consequence of Mendel's work.

What is undeniable is Darwin's acute intuition. Without the help of these later disciplines, Darwin could sense the existence of laws, as he says, unknown to us, which rule the evolution of living beings, and hence the possibility that natural selection could work in perfecting the different species.

One of several objections pointed out by some scientists is that Darwin did not solve the existence of the eye as a product of evolution. This question has not been solved yet. Many different hypotheses have been exposed, but much more has to be investigated before reaching a complete explanation. They also insist that many aspects of Darwin's work had not been adequately developed and that many aspects of it remain absolutely without explanation. Let us be honest: Darwin could not use the laws of heredity, the genes, or their sequence in DNA to explain every aspect in evolution. Darwin answers to this situation in this way (page 64): *"Several writers have misapprehended or objected to the term Natural Selection. Some have even imagined that natural selection induces variability, whereas it implies only the preservation of such variations* [that] *arise and are beneficial to the being under its conditions of live. No one objects to agriculturists speaking of the potent effects of man's selection; and in this case the individual differences given by nature, which man for some objects selects, must of necessity first occur. Others have objected that the term selection implies conscious choice in the animals which become modified; and it has been urged that, as plants have no volition, natural selection is not applicable to them! In the literal sense of the word, no doubt, natural selection is a false term; but who ever objected to chemists speaking of the elective affinities of the various elements? And yet an acid cannot strictly be said to elect the base which it in preference combines. It has been said that I speak of natural selection as an active power or Deity; but who objects to an author speaking of the attraction of gravity as ruling the movements of the planets? Everyone knows what is meant and is implied by such metaphorical expressions; and they are almost necessary for brevity. So again it is difficult to avoid personifying the word nature; but I mean by nature, only the aggregate action and product of natural laws, and by laws, the sequence of events as ascertained by us. With a little familiarity such superficial objections will be forgotten."*

Let us be objective. The fact that many points were not explained by Darwin's original work does not mean that the idea was without merit. Christopher Columbus, the discoverer of America (notice that I say the *discoverer*, not the first settler), received recognition for making known a whole continent, a very large part of the world, even if he did not explore all the different regions of this immense piece of land. He even died ignorant that he had discovered a new world. But does that diminish the importance of his exploit? No. Evidently in a man's life there is not enough time to explore a field as large as science piece by piece. Even today, after 500 years, all of America has not been fully explored. There are many regions that we know exist, but have not been studied thoroughly. If in 500 years there has not been enough time for full-scale research of a continent, is it a wonder that in a few years of Darwin's life, all aspects of the evolution of millions of species had not been meticulously studied and answered properly? Darwin was a pioneer in a completely unknown field, as Copernicus was in astronomy. Both were pioneers, but neither of them could conduct exhaustive research in their fields. Even today we are adding new findings in every field of study. The more we research and understand, the more questions that arise, compelling us to look for answers.

Darwin was not the first to propose evolutionary theories. Other scientists, including his grandfather, Erasmus Darwin, had proposed evolutionary theories. Darwin himself explains the situation in his own words (page 3): *"Lamarck was the first man whose conclusions on the subject excited much attention. This justly celebrated naturalist first published his views in 1801; he much enlarged them in 1809 in his Philosophie Zoologique, and subsequently, in 1815, in the introduction to his Historie Naturel des Animaux sans Vertebres. In these works he upholds the doctrine that all species, including man, are descended from other species. He first did the eminent service of arousing attention to the probability of all change in the organic, as well as in the inorganic, world being the result of law, and not a miraculous interposition.*

Geoffroy Saint-Hilare, as is stated in his Life, written by his son, suspected as early as 1795, that what we call species are various degenerations of the same type. In 1828 he published his conviction that the same forms have not been perpetuated since the origin of things. Dr. W. C. Wells read in 1813 before the Royal Society an article in which he recognized the principle of natural selection. The Hon. And Rev. W. Herbert, afterwards Dean of Manchester, in the fourth volume of the Horticultural Transaction, 1822, and in his work on the Amaryllidaciea (1837, pp. 19, 339), declares that 'horticultural experiments have established, beyond the possibility of refutation, that botanical species are only a higher and more permanent class of varieties.' He extends the same view to animals. The dean believes that single species of each genus were created in an originally highly plastic condition, and that these have produced, chiefly by intercrossing, but likewise by variation, all our existing species. In 1826 Professor Grant, in the concluding paragraph [of] *his well-known paper (Edinburgh Philosophical Journal), vol. xiv, p. 283 on the Spongilla, clearly declares his belief that species are descended from other species, and that they become improved in the course of modification."*

Darwin became famous from his work because he was the first to collect factual evidence for it. During his five-year exploratory voyage aboard the *H.M.S. Beagle*, along the coast of South America, the Galapagos Islands, and other islands in the Pacific, he observed the proposed theories *in situ* of nature at work. Some species that had originated with a common ancestor developed different characteristics once they became separated geographically from other members of the same species. Until this point, his theory was not completely disputed. Some renowned scientists such as Thomas Henry Huxley, Sir Charles Lyell, and Asa Gray supported Darwin's work. Some debate originated with this publication, but actually the increase of opposition to his theory started when Darwin published his book *The Descent of Man* (1871). In this work, Darwin outlined his theory that man came from the same group of animals known as apes, such as the chimpanzee, gorilla, gibbon, and orangutan. Here is where the storm broke, supported mostly by religious groups.

It is perfectly understandable that an idea that was against the teachings found in the first book of the Bible, *Genesis*, where it is clearly stated that the human being was the product of God's special creation, had to face a strong opposition. After hundreds and hundreds of years of teaching that man was created by Divine intervention and that he was given the privilege of dominion over all other animals on Earth, the idea that we had evolved from a branch of the animal kingdom was unacceptable. It appears to be an abandonment of our unique position as superior living beings selected by Divinity to rule the Earth. The opposition became more acute when the critics of Darwin's theory, without reflecting on it and in an effort to ridicule it, spread the idea that Darwin supported the point of view that human beings descended from monkeys. Anthropoids are remarkable human-like primates an attribute that should not be embarrassing, but endearing. This was a misinterpretation of the reality exposed in *The Descent of Man.* Charles Darwin says in the first chapter of this book (page 395): *"He who wishes to decide whether man is the modified descendant of some preexisting form, would first enquire whether man varies, however slightly, in bodily structure and in mental faculties; and if so, whether the variations are transmitted to his offspring in accordance with the laws which prevail with the lower animals. Again, are the variations the result, as far as our ignorance permits us to judge, of the same general causes, and are they governed by the same general laws, as in the case of other organisms; for instance, by correlation, the inherited effects of use and disuse, etc.? Is man subject to similar malconformations, the result of arrested development, of reduplication of parts, etc., and does he display in any of his anomalies reversion to some former and ancient type of structure? It might also naturally be inquired whether man, like so many other animals, has given rise to varieties and sub-races, differing but slightly from each other, or to races differing so much that they must be classed as doubtful species. How are such races distributed over the world, and how, when crossed, do they react on each other in the first and succeeding generations? And so with many other points."*

Most of Darwin's critics have never read his works. They condemn them without knowing the facts or the reasons. Emotions run high when an idea opposes some religious dogma. But . . . the Earth moves!

Note

1. Gordon, *De l'Espèce,* 1859, volume. ii, livre 3. Quatrefages, *Unité de l'spèce Humaine,* 1861. Also lectures on anthropology, given in the *Revue des Cours Scientifiques*, 1866-1868.

III

Darwin's First Ommission: The Little Developed Human Newborn

Apart from other characteristics not completely explained by Darwin in relation to the human being, the fact that the state of development of the newborn is very different from that of the other primates should suggest that something else should be considered before connecting all those species. Darwin did not realize the outstanding difference between human newborns and the other apes' newborns. Darwin made a big mistake when he said (page 397): *"The whole process of the most important function, the reproduction of the species, is strikingly the same in all animals, from the first act of courtship by the male, to the birth and nurturing of the young. Monkeys are born in almost as helpless a condition as our own infants. . . ."*

I emphatically can say, that that is not so.

Darwin has demonstrated to be a keen observer, taking into account the most subtle details and studying them until reaching surprising conclusions. His cunning eye captured even the smallest difference in detail that, for most people, would be impossible to notice or to realize the important connection or consequence of its existence.

The fact that he was such an acute observer is what makes astonishing the fact that he overlooked one of the most outstanding differences between man and the other primates. When he says: *"Monkeys are born in almost as helpless a condition as our own infants"* he makes a statement that does not correspond to reality. Everyone can make a slip, but this is too notorious. Let me explain why I disagree profoundly with Darwin's assertion.

The state of development of the newborn of the great apes and the newborn of the other primates is very similar. All primates, from the prosimians through the lemurs, the monkeys to the great apes or anthropoids, have a common characteristic not shared by the human being. The newborns of all the primates are born in a very developed state. Not so with the human baby. Man's newborn is in an advanced stage of development at birth if it is compared to a marsupial, but is little developed if we compare him with such animals as the newborn rhesus monkey, baboon, chimpanzee, or gorilla.

The newborn orangutan, gorilla, chimpanzee, gibbon, and all the other primates are born in an alert, advanced stage of development with coordinated reflexes. Human newborns are totally helpless when compared to these animals. The human newborn has no coordinated reflexes. He cannot grasp anything; even if someone puts his finger in his hand, the baby does not react to it. It takes several days until he can grasp a finger. The human newborn cannot see. For several weeks he does not react to light, and takes a month or longer to start reacting to images. The newborn cannot hold his own weight, and cannot move his body or roll from one side to the other. The parents have to move him and accommodate him frequently. On the other hand, all primates, including the great apes, which are considered our closest relatives, are born with sight and coordinated reflexes, such as the reflexes to grasp the mother's fur. They can move, roll, and adopt a comfortable position when needed. They can hold their own weight without any help from the mother. This information based on concrete facts, seems to point to the possibility that these two groups, man and anthropoid, have taken two different paths of evolution and as a consequence, they have occupied distinct ecological niches. The anthropoid group is more adapted to immediate physical action and the human group is more adapted to an elaborate preparation emphasizing intellect and reason. Man's adaptation points to a role in which the cerebral development will be far more important than the physical one.

The human newborn has a brain weight of approximately 350 grams, which is almost the weight of an adult chimpanzee's brain. This signifies that man's development is not balanced, the physical aspects are underdeveloped and at a disadvantage, while the cerebrum, although not completely functional, is still in an advanced or advantaged stage of development (page 15).1

The human fetus demonstrates involuntary reactions at a very early age. At eight weeks, the human fetus can turn its head, and at sixteen weeks he can suck. These are reflexes that do not correspond to mental control or coordination, but are involuntary and possibly are the beginning of reflexes that the anthropoids will later develop most (page 15).2

The human fetus and the human infant develop very slowly. Human infants need long periods of infancy to acquire the necessary agility to stand up, walk, run, and use the hands voluntarily, and they need many years to be able to function independently of their parents.

One would think that from the survival point of view, the human infant does not have many advantages. On the other hand, it is an obvious advantage that the anthropoid baby is quickly able to react and take care of himself, even if it also needs an expanded infancy. This would apparently be a positive characteristic for the preservation of the species (page 15).3

In evolutionary terms, what does this retarded state of physical development in man's offspring mean? Does this state have some advantages for the survival of the species? It would have to, because otherwise the species would not have preserved itself. Later in this work, we will see how important this retarded state of the newborn is and the enormous consequences that this fact has implied (page 15).4

Before developing the explanation to this question, first it will be necessary to call attention to an important human trait: the lack of specialization. When an animal finds itself occupying a very specialized niche on the evolutionary scale, it also finds itself in a precarious position because the slightest imbalance can cause his extinction. If this animal depends on one type of food, the minute that this food disappears because of changes in the climate, an epidemic, or a natural catastrophe, this animal will perish and eventually so will the species (page 15).5 Remember what happened in China a few years ago. Bamboo is a giant grass that blooms in the last stage of its life. After blooming and seeding, the plant dies. When that happened, the giant panda that feeds exclusively on bamboo began to perish by starvation. Apart from poaching, this was a major ingredient that put this animal on the list of endangered species.

If an animal's specialization is related to climatic conditions, as soon as the climate changes significantly, this would also mean the end of the species. A good example is the fossils of tropical plants and animals found in Antarctica.

Specialized organisms are extremely efficient as long as the conditions that gave rise to their speciality are present, but at the same time, they are extremely vulnerable to any change in the required conditions. The specialized organisms have abandoned more diverse alternatives for survival to ensure a more efficient survival within their special living conditions (page 15).6

Specialized organisms contrast noticeably with generalized organisms. The generalized organisms are not subject to a rigid set of conditions and are capable of adapting and being flexible in a number of environments. Evolution continually produces both types of organisms (page 15).7

From the evolutionary point of view, human beings are generalized organisms, at least in appearance. They adapt to any climate, they are capable of eating many kinds of plants and animals, they adapt to any altitude (they can live at sea level or in the highest points of the Himalayas), they can live at any latitude (at the North or South Poles or at the Equator), and they adapt to very diverse circumstances and conditions and always survive as a species (page 16).8

Human beings have achieved great success in survival as a species due to lack of specialization. Another animal that has survived due to this lack of specialization is the common rat (*Rattus norvegicus*). This animal is capable of eating almost anything, and living in any climate, and is able to adapt to almost any condition, because of this lack of specialization it is a very successful species (page 16).9

One could think that man owes his success as a species to the specialization of not having a specialization. This apparent lack of specialization is somewhat superficial. This point will be developed later in this work (page 16).10

Historians, anthropologists, philosophers, and thinkers in general have made many efforts to decipher the origin of civilization. Unfortunately, they started their research halfway, and not at the beginning, or at a more remote point of origin. This starting point of civilization was the condition of the human newborn that obliged man to be bipedal, and has been the reason and cause of the development of civilization, as we will see throughout this work.

Notes

1. Man, the Paradoxical Primate, Jordi Fuentes, 1985.
2. Ibid.
3. Ibid.
4. Ibid.
5. Ibid.
6. Ibid.
7. Ibid.
8. Ibid.
9. Ibid.
10. Ibid.

IV

Man, The Only Nidicolous Primate

Man is physically a very unspecialized animal; he doesn't possess the lion's fangs, the speed of the gazelle, the strength of the buffalo, the serpent's venom, the agility of the squirrel, nor the ability to escape flying or swimming rapidly from danger. Man is not specialized offensively or defensively. Man's survival has depended strictly upon his social organization or on his ability to cooperate and live with others of his kind. It is in this group, and this group may be a family, a tribe, or a clan, where individuals have found their defense and survival. In such a group, when food is found, whether it be by gathering, hunting, etc., once it is found by anyone, the rest of the group would most likely benefit by it (page 17).1 The fact that man is a gregarious animal living together with other men is quite related to the fact that he is born in a little-developed stage (page 17).2 Darwin stated this fact, but he did not realize the importance of it, losing the opportunity to develop and show the tremendous importance of this condition that is the origin of our civilization. Darwin said (page 443): *"It has often been objected to such views as the foregoing, that man is one of the most helpless and defenseless creatures in the world; and that during his early and less well-developed condition, he would have been still more helpless."*

The relative helplessness that man faces when he is born obliges him to live within a group. The mother has to hold the newborn, as he cannot cling to her as do anthropoids. The newborn does not have the strength to hold his own body weight, and the mother has no hair to which the baby can cling, the way the primates do. She, however, cannot have him continually in her arms, as she has other obligations to attend to, such as searching for food and taking care of other older children and her mate. All of this taken into account, along with the fact the newborns need to sleep as much as twenty one hours a day during the first days of their life, obliges the mother to find some place to leave the baby for periods of time to enable her to free her hands. We will call this place "the nesting place," and it may be anywhere between two rocks, under a bush, a hiding place on the plains or in a cave. Wherever it is, there is no doubt that the mother needs it (page 17).3

This nesting place is the origin of the human group. Man's activity and independence and his need for a nesting place make him, paradoxically, the only nidicolous anthropoid. *"Nidicolous"* means that he needs a nest (*"nidus"* in Latin). Men are completely diurnal animals, and at night they need this nesting place to sleep and as shelter against predators. Their activity ceases with sunset (page 17).4 In the most primitive hunting and gathering groups today, in spite of the fact that they may be considered nomadic, they still need a shelter, a bivouac, while they search for food and where they may rest and be protected from predators (page 17).5

This group that the helpless infant forced man to live in could have been, in some instances, as small as mother, father, and child. Of course, the larger the group, the more advantages it offers, just as a larger nation usually offers more advantages over a smaller one (page 17).6

If a man needed a nesting place, this nesting place must have been on the ground. This statement may initially appear unlikely, as even gorillas usually make nests in trees in order to sleep. Gorillas, however, as they grow older and heavier, have a tendency to make their nests less high in trees, and in many cases they make their nests on the ground. Logic would indicate that human beings were not arboreal (as I will explain in chapter VII); hence, they would make their nesting places on the ground (page 17).7 They obviously could not make a nest for a helpless newborn in a tree. Can you visualize a sort of crib with a helpless baby on the branches of trees? Only birds do that. The idea does not make any sense. Under those special conditions, men most likely made their nest on the ground. This nest could be made more comfortable with the use of leaves, grass, and animal pelts. People have most likely noticed that their pet dog circle several times before lying down. This is an atavism of times in their past when they needed to make the ground softer, which they achieved by stepping on the grass or wherever they had chosen to lie down. Our ancestors most likely also prepared a nest, especially when they knew they could make use of it for several days at a time (page 17).8

Now we are beginning to develop a picture of the hominids; they were terrestrial animals (not arboreal), that walked erect on two feet (see the chapter on The Upright Man), and they lived and traveled in a group. Anthropologist Mary Leakey, while working in Laetoli in Tanzania, found some well-preserved fossilized footprints of a group of at least three hominids, one appearing to be a child. These footprints are approximately four million years old, and the group traveled together such as we might today (page 17).9

All animals' behavior and activities are governed by the fact that they must reproduce themselves in order for the species to survive. Biologists' and ethologists' studies have shown that reproduction is the greatest force in animals' lives. It is necessary that they reproduce themselves in order to perpetuate the species, and it is important to point out that it is the species that is important here, not the individual. The salmon is a good example of how compelling the need to reproduce the species can be. The salmon, after living in the ocean where food is abundant, reaches sexual maturity and abandons the ocean to begin its long, arduous trip to its birthplace. To reach its birthplace or "nest", it must overcome many obstacles, including swimming upstream, and in the course of this voyage many salmon perish (page 18).10

The different behaviors that we observe in animals, whether they be social animals or solitary animals, are governed first by the need to reproduce the species and second by the need to find food without which the animal would perish. The need to reproduce oneself is not only manifested in the act of mating but also in the caring for and protecting of the young until the offspring is able to fend for himself. We can only call reproduction successful when the offspring has reached the stage where he may now independently care for himself. Mating and giving birth many times is not sufficient to ensure reproduction of the species. Of course, there are many cases where parental intervention ceases upon giving birth or upon laying an egg, but even in these cases where apparently there is a total lack of parental care, if we look harder, it is evident that the parents take considerable care in selecting the most favorable places, conditions, and times of year to guarantee the survival of their progeny. The migration of the California gray whale and the migration of many species of birds are only two of the many examples of this need to find the most favorable sites for the raising of offspring (page 18).11

Human beings, of course, have the same compelling need to reproduce themselves and preserve the species by protecting and caring for their young. Since the human offspring requires very specific conditions to be able to develop to a stage where he may take care of himself, his parents are necessarily subjugated to these conditions. Their everyday actions, perhaps on a subconscious level, are governed by these needs and requirements (page 18).12

When an astronaut, after making a trip to the moon, later returns to Mother Earth and goes home, he is acting according to the same compelling force that makes the bushman return to his home base after having completed a hunting expedition. The office employee and the crew of an airplane having returned also demonstrate the need to "go home." The well known saying "A man's home is his castle" reflects the popular awareness of this need to have a special place where his family resides and where he may return after completing work (page 18).13

Without the need for a nesting place, civilization would not exist. The grouping of nesting places formed communities that later developed into villages, towns, and cities. Without these communities, the development of the different specialties that created the variety of trades according to the development of different skills would be impossible. Governments would not exist because they are the result of a solution needed to regulate the life in communities, either cities or countries. In order to regulate the relation between countries, the United Nations was created.

All activity of the present human race is an immediate consequence of the human being state as a nidicolous animal. Should we be like the other anthropoids without the need for a nesting place, civilization would be impossible and we would still be like the other anthropoids, organized in a family group with no relation to other families, as happens with chimpanzees, gorillas, and orangutans. Neither Darwin nor anybody else has realized the importance of the helpless state of the human newborn. It is possible that some objections arise to this conclusion for instance, the fact that other nidicolous animals have not developed civilizations. The answer to this objection is: No other nidicolous animal was upright and bipedal, with the possibility to develop very functional hands. They continue being quadrupedal and the four limbs are legs, not arms and hands. Conclusion:

The entire organization of our modern-day society is based on the fact that man is the only nidicolous primate. The human nesting place is the cornerstone of all human society regardless of race, geographical location or culture. This nesting place has given rise to present-day civilization in which thousands of nesting places have created modern cities with all their structural and organizational complications. Should human beings be like the other primates, that is to say, not nidicolous animals, civilization would not exist.

Notes

1. Man. the Paradoxical Primate, Jordi Fuentes, 1985.
2. Ibid.
3. Ibid.
4. Ibid.
5. Ibid.
6. Ibid.
7. Ibid.
8. Ibid.
9. Ibid.
10. Ibid.
11. Ibid.
12. Ibid.
13. Ibid.

V

Man, a Territorial Animal

Nature has prescribed that man is a nidicolous creature in that he needs a nest to take care of the newborn. The newborn is devoid of reflexes that would allow him to protect, shelter, or hide himself. Neither does he have those instincts that facilitate the search for food, such as hunting and gathering. This need for a nest is so imperative that human beings become guardians and defenders of the place where the family is harbored. The fact that humans are newborn nidicolous animals means they must defend their home or territory, whatever its size and extension. Man becomes, due to the conditions of his birth, a territorial animal par excellence.

In order to defend his territory, man is obligated to develop his aggression toward any other animal or human who represents a danger to the well-being of his family. Aggression originates as a consequence of this need. The well-being of his family implies not only the physical protection of his home, but includes the territory where he finds sustenance for his family. The fewer sources of food existing in a territory, the larger the area man must protect in order to guarantee enough food for his clan. In our society with urban lifestyles, man continues to display the same original characteristics; he has to have a nesting place, a place to raise his family, and he is obligated to defend his territory against any intrusion. In our society the nesting place is the home, and the feeding territory is our place of employment. The need for these two conditions is the origin of the highly developed sense of possession for property found in human beings.

Territoriality, the source of a sense of ownership of property, is highly developed in human beings. We all feel we own something. If we own real estate, we are the absolute owners of it, and if we own other belongings, even if not real estate, we still consider ourselves owners, and no one may borrow our belongings without our express permission.

The development of the sense of property born as a consequence of the condition of the human newborn is extended to all those articles and household goods that serve a particular purpose in the family group. The primary importance represented by these objects is that they bring about an immediate sense of material well-being, such as occurs in relation to a tool, a kitchen utensil or weapons for hunting and defense. This sense of well-being can also be extended according to the mental relationship that the individual human attaches to objects that don't have a practical value but merely produce a pleasing sensation upon contemplation. This is the origin of artistic sensibility and is the reason for the birth of the speciality we call art.

"The American Dream" to own your own house is the dream of the vast majority of populations, not only the population of the United States. Owning one's own property is the guarantee of physical security of the family. Upon obtaining his own nesting place, the human being becomes a territorial proprietor. His property may be only a cave, a hut, an igloo, an apartment within a building, an isolated house surrounded or not by land, a mansion, a farm, a ranch, or otherwise. Whatever its size, "a man's home is his castle": In other words, his house is the territory that he owns and over which only he has jurisdiction and dominion.

Some animals are territorial in defense of both the nesting place and the minimum territory necessary to find enough food for the survival of the family. The poorer in food resources an area is, the larger the area must be, and the greater the effort of territorial defense to guarantee its exclusive use. This effort usually is expressed as an attack on the trespasser.

Territoriality in most animals is seasonal, although there are cases of permanent territoriality during the life span of the animal.

What are the implications that favor this territoriality trait in the human being? All expressions of human life, dating from the most remote traces of existence millions of years ago, are derived from the characteristic of territoriality, which is expressed in every activity, culture, or human group. Man is a gregarious or social being, tending to form groups. The first groups are among those belonging to the same family, or related by marriage. Upon marrying into other groups, immediate family ties begin to weaken, but even so they remain related. Little by little, through his expansion fostered by the sexual interaction of the individuals of the group, the group continues to grow, forming villages. These later, upon acquiring more members, become towns and then cities.

The fact that humans are territorial and social animals creates a complicated relationship between members of the group, thus giving rise to social organization. This organization allows the group to develop its natural functions of living and procreation, all within the frame of the physical conditions of the territory where the members live. The influence of territorial conditions and the social imperative gives rise to infinite variations of behavior that when passed down from generation to generation, become part of what we call culture. Without social organization with strict rules of behavior, it would be impossible for the individuals of the group to live together. If they behaved only in response to their territorial imperatives, living together would be impossible. These norms of conduct guarantee the right to territorial property even when the property is assembled in villages, towns, and cities.

The different climatic conditions that exist on Earth create great differences between geographical zones, giving rise to the need for special adaptation in order to survive and prosper under each different condition. Each culture develops in accordance with these characteristics, focusing on the life-enhancing qualities from each climatic zone.

Nomadic cultures exist whose ways of life are dependent on the seasons and on the wild and domestic animals that provide them food. They must move from place to place, following the annual cycle in search of food. These nomadic cultures have adapted to the deserts, Arctic zones, tropical jungles, and marine environments. Their nomadic life forces them to bring their nesting places with them, and they defend their territories according to the amount of food found there.

It has been easier for human beings to obtain food since inhabiting cities, living in a group of houses or "nests," united under a collective organization, and supported by industrial development. The food is not found directly in the territory, but rather through the system of buying. The individual defends his territory of supply, in this case, through employment, which in turn guarantees him the ability to obtain the product (money) and make the exchange (buy) in the supermarket. The defense of territory is replaced by the defense of job. Man is obligated to defend his position at work against any intrusion (competition with other workers) in order to guarantee his ability too feed his family. Instead of defending an area in the woods, jungle or in the desert, man then defends his job. We will use the word "job" instead of "work" because of the parasitic working relationships, or persons who are employed but really do not work, obtaining their gratification based on the work of others, becoming society's parasites.

We can detect the beginnings of the defense of employment in the efforts to obtain a better education, whether it be a trade or a college career. In both cases, the objective is to obtain a tool that will permit the individual to struggle for an advantage in his competition with others, to obtain a better job. This struggle is what has given rise to the saying "the struggle for survival," which describes the continuous effort to obtain a better job in competition with others, and consequently a better economic life for the family. This struggle occurs in order to protect the territorial zone where the food for the well-being of the family is found.

The majority of human activities originate from this innate need obtain or defend a territory.

VI

Human Aggression

There have been many attempts to explain human aggression by the use of reason, but unfortunately most explanations do not consider the biological source for this behavior. Most explanations of human aggression try to relieve humanity of the responsibility that corresponds to a behavior so inhuman as is aggression directed to one's own species taken to the extreme of a total destruction of individuals, and even complete nations (page 52).1

In order to find the origin of this erratic and apparently illogical behavior, it is necessary to consider the biological factors that most likely have influenced its development, once again leading us to the connection between biology and anthropology. In this hopefully logical analysis, we will start with the general biological factors common to the appearance of aggression in all animals in order to build a sturdy foundation for a theory (page 52).2

Aggression in animals is normally exhibited as a consequence of two types of stimuli. The first stimulus is the necessity to find nourishment, and the second is the need to procreate. We will study each of these stimuli separately (page 52).3

The need to obtain nourishment may, in turn, be divided into two different categories: the predator in search of prey and the herbivores (there are animals that do not belong to either category, such as animals that feed primarily on carrion) (page 52).4

In predators, aggression originates from the immediate need to eat. Certain animals are their nourishment, and they have no choice but to hunt these animals down. Aggression is innate in predators; it is part of their nature or a behavioral habit. If aggression were not stimulated by hunger, they would perish. Once the predators' hunger is satisfied, they lose their aggressiveness and may become totally indifferent to those animals that are normally their prey (page 52).5

Herbivores are normally not aggressive toward animals of the same species unless they are sedentary species where the obtainment of food is directly related to a certain area or territory, which they defend when new individuals are introduced in the area. The defense of this territory is necessary to ensure sustenance (page 52).6

The need to procreate is the second stimulus causing animals to exhibit aggression. The kinds of situations that bring about aggression relating to reproduction are threefold: the need for the pair to reproduce, the necessity to establish a territory in which to raise the young, and the defense of the young until they are able to fend for themselves (page 52).7

The need for a pair to reproduce causes aggression directed against the same animal species. The necessity to establish a territory in which to raise and defend the young causes aggression against the same or different species. Aggression is manifested against potential enemies, and even an animal's own offspring may be the object of aggression if the offspring has reached an age where he may be considered a rival (page 52).8

With certain animals that live in a highly organized group, such as the wolf, the leader of the group is that wolf which is the most aggressive. This fact alone would lead one to believe that the goal of aggression is to become the dominant member of the group. In reality, the goal of the wolf's aggression is to win the right to perpetuate (unconsciously) its genes, something done instinctively, not as a product of reasoning, as they do not know that there are genes to be transmitted. In a wolf pack, only the dominant pair reproduce. This is the mechanism that regulates the number of predators and maintains the balance between predator and prey. If all predators would reproduce, soon the scarcity of food would cause the species to perish. Although on a somewhat different scale, the same can be said for baboons; but sometimes baboon females that are in estrus will mate with subordinate males (page 52).9

It is very common when writing to commit some mistakes. Sometimes those errors are not important, as they do not carry any consequence, but if the mistake has some connection with some other ideas or realities, then the slip can be detrimental to the acceptance of the idea, because the reader loses confidence in the author. This situation is illustrated in Darwin's long exposition about certain facts in relation to the populations of animals and human beings in the chapter entitled *"Rate of Increase."* Darwin says (page 430): *"If we look back to an extremely remote epoch, before man had arrived at the dignity of manhood, he would have been guided more by instinct and less by reason than are the lowest savages at the present time. Our early semi-human progenitors would not have practiced infanticide or polyandry; for the instincts of the lower animals are never so perverted* 10 *as to lead them regularly to destroy their own offspring, or to be quite devoid of jealousy."*

At this point we have to reflect about these ideas. First, it is very risky to say that our semi-human progenitors would not have practiced infanticide or polyandry. We have no information about this assumption. But if we have to arrive at conclusions comparing the behavior of our ancestors with our not so "perverted" society, I am forced to say that in our present-day society, in the twentieth century, infanticide, mostly female infanticide, has been, and is, practiced as a routing in some Asian cultures. The reason apparently is the elimination of mouths to be fed, considered not suited for the production of food for the family group. In ancient times, in cultures as developed as the Spartan culture, infanticide was common, mostly for the elimination of defective newborns. In the Andean cultures of South America, we find ritual infanticide as a religious ceremony. The Inca empire, one of the most advanced civilizations of South America, practiced this kind of ritual. There are many examples in Peru and Chile, where the naturally mummified bodies of boys and girls have been found, buried alive at very high elevations. With regard to polyandry, I don't see anything "perverted" in this practice. In some very poor cultures in Asia, it is practiced as a means to provide for the sexual life of a group of brothers, because females are scarce. Poverty forces these people to eliminate females, and their scarcity forces some brothers to marry and share a woman, as they cannot maintain one wife each. There are many species of animals that practice polyandry as a normal way of reproduction, and this practice does not presume the existence of a "perverted" mentality. With regard to the destruction of their own offspring, we have plenty of evidence that male bears routinely kill any cub they can reach. There is no reliable information about the reasons for this behavior. With lions, we know that males kill all the cubs that do not belong to them, such as when a new male takes over a pride without a male, or after he has defeated the dominant male. Soon after female lions lose their cubs, they go into estrus, and the new male mates with them.

Darwin continues developing this idea (page 430): *"There would have been no prudential restraint from marriage, and the sexes would have freely united at an early age."* That is not only possible, but very probable, as I have explained in the chapter *Man, a Case of Neoteny*, as the presence of neoteny allowed the survival of the hominid species. Darwin continues (page 430): *"Hence the progenitors of man would have tended to increase rapidly; but checks of some kind, either periodical or constant, must have kept down their numbers, even more severely than with existing savages. What the precise nature of these checks were, we cannot say, any more than with most other animals. We know that horses and cattle, which are not extremely prolific animals, when first turned loose in South America, increased at an enormous rate. The elephant, the lowest breeder of all known animals, would in a few thousand years stock the whole world. The increase of every species of monkey must be checked by some means; but not, as Brehm remarks, by the attacks of beasts of prey. No one will assume that the actual power of reproduction in the wild horses and cattle in America was, at first, in any sensible degree increased; or that, as each district became fully stocked, this same power was diminished. No doubt in this case, and in all others, many checks concur, and different checks under different circumstances; periodical deaths, depending on unfavorable seasons, being probably the most important of all. So it will have been with the progenitors of man."*

Here, Darwin does not consider the predatory action of carnivore animals as the principal means of controlling the population of the different species. Predators are the main force that controls the expansion of any species. This has been unfortunately proved by the introduction of a species to a new territory where natural predators for the species do not exist; the new arrivals run amok without control, detrimental to other species and the ecosystem of the area or country. Examples are the introduction of rabbits and the toad (*Bufo marinis*) in Australia, and the introduction of opossums in New Zealand. In the United States we have the example of the introduction of starlings and English sparrows from Europe that are destroying native species such as the woodpecker. When the new arrivals compete with them, and as they are more vigorous, the native species are forced to abandon their nest to the foreign species. Cattle and horses in South America, as horses and donkeys in the deserts of southwestern United States, reproduced in great numbers because they had no predators in their habitat that could control the number of survivors. Darwin considers the "unfavorable seasons" as a probable cause of limiting the number of individuals of a given species, but does not take into account either the predators or the diseases.

Cannibalism is an expression of aggression with different origins according to species. Frogs are cannibalistic because they do not distinguish shapes visually. They react to movement. If the object that moves in front of them is another frog, they eat it. Other species are cannibalistic to suppress rivals, or as a trait of dominance. Human beings have shown cannibalistic behavior corresponding to several and different ideas. In some cultures, cannibalism is considered a way of dominating the enemy and absorbing his physical properties. In some other cultures, cannibalism was only a way of procuring protein for their diet. In Easter Island, cannibalism was one of the causes of wars between the two major tribes inhabiting this small island on which no large animals were found. Fishing and hunting enemies provided the only sources of protein. Once these enemies were captured alive, they broke their legs in order to prevent their escape. They were maintained alive and consumed little by little during a large span of days or weeks. Fellow men of the other tribe were only considered to be a food source.

Now that we have made this small inroad into animal aggression, we will attempt to analyze human aggression.

Humans, like all other animals, exhibit aggression in response to the two needs mentioned before: to obtain nourishment and to procreate. Hunger is what forces the animal to obtain food, which is accomplished by killing other animals. To ensure the source of food, they show aggression toward animals of the same or different species who enter their hunting or collecting territory. Aggression is present also, in response to the need to procreate.

Up to this point, it appears that man does not depart from the norm of typical animal behavior, and his extraordinary aggression toward his own species and other animals (potentially refined to the point of total destruction of life on Earth) would remain unexplained. As was mentioned earlier, even normally passive species, such as herbivores, exhibit a high degree of aggression toward their own kind when they are responding to the need to reproduce. Normally, this need occurs annually; aggression becomes more acute the closer mating season gets, and does not diminish until after mating season is over. After having paired off and mated, biological pressures are alleviated and aggression disappears to the point that males that were enemies during the mating season return to share the same groups and live tranquilly together for the rest of the year until the hormonal activity in their systems compels them to once again seek out the females of their species.

In human beings, the possibility of mating does not limit itself to once a year, but rather continues during the entire year. This means that the human being is continuously influenced by a high hormonal activity that translates into aggressive behavior. In other words, the need to compete for a female, to mate and relieve biological pressures and urges, and the continuous need to protect and feed the offspring maintain man in a chronic state of alertness and aggressiveness (page 53).11

Human beings are able to give birth to an offspring once a year. This offspring, however, requires many years of parental attention before he is ready to be independent, due to the little-development state in which he is born. Therefore, the parents are obliged to defend and take care of the several offspring during many long and different stages of development. The continuous pressure on the parents to provide food and protection remains for years. Aggressiveness that in other animals is temporary is the behavioral norm for human beings (page 53).12 We are chronically aggressive.

There are other ingredients that make aggression a norm in human behavior. One ingredient is the *"cerebralization"* of man, by which man has developed technique. Cerebralization allows man to give profound consideration to any idea he wishes to develop in achieving his objective. His objective, in this case is aggression, not only against those animals that constitute his diet but also against individuals of his own species. Rivalry to obtain the best territory and the best natural resources becomes more acute (page 53)13 as the possibilities of success increase. More and more resources and energy are used to achieve this objective, using the help provided by a more developed brain.

These would be the biological roots of human aggression, but there are cultural influences that cause aggression to grow and perpetuate itself. Social value is given to aggression, converting it into a desirable quality worthy of respect and dignity. Sometimes aggression may have ideological and religious meanings, such as with head-hunting among the Jibaro Indians of Ecuador and New Guinea. Cannibalism is another expression of violence imposed by culture, sometimes considered as a confirmation of the destruction of the enemy and other times as a method of acquiring the virtues or powers of the vanquished (page 53).14

Violence and aggression have been at the service of religion many times throughout history. With religion, aggression's existence has been justified and aggression has been idealized, such as in the human sacrifices of the Mayan culture, religious wars such as the Crusades, the killing of unfaithful Indians in the American Colonies, the infamous "autos-da-Fé" of the Holy Office of the Catholic Church, and the persecution and killing of Jews in many countries in the world, such as took place in Russia and Germany under the command of the Third Reich (page 53). 1 This list of man's aggression against man is inexhaustible (page 53).16

Other times, aggression is manifested due to man's zeal for profit and gain, such as happened during the period of slave traffic, or the annihilation of the American Indians in order to colonize coveted land. There are many cases in which economical motives are the cause of aggression. The aggression that began in man's ancestors as a consequence of a chronic reproductive cycle now is manifested on a national scale, with one country attacking another under any pretext (page 53).17

Once again, the human being offers a paradox. His survival depends upon the group and its organization; therefore, he is considered a social animal. The paradox is that as a social animal, he manifests supremely antisocial behavior, such as aggression, against his own kind. The tolerance shown toward his own kind, which has been the base of his social organization and has allowed his expansion and success as a species, is becoming inhibited as he increases his *cerebralization.* With *cerebralization*, personal egoism is also developing to the point that man uses his kind when he needs them but will destroy them when he considers them an obstacle to the achievement of his goals. This is indeed a paradox (page 54).18

On this subject Darwin says (page 471): *"I FULLY subscribe to the judgment of those writers who maintain that of all the differences between man and the lower animals, the moral sense or conscience is by far the most important. This sense, as Mackintosh* 19 *remarks, 'has a rightful supremacy over every other principle of human action.' It is summed up in that short but imperious word ought, so full of high significance. It is the most noble of all attributes of man, leading him without a moment of hesitation to risk his life for that of a fellow creature; or after due deliberation, impelled by the deep feeling of right duty, to sacrifice it in some great cause."*

I profoundly disagree with this rather poetic and romantic judgment about human beings. Let us consider this point: How much money is spent in helping the needy of our species? How much money is spent in trying to destroy our fellow members of the same species? The disproportion is in favor of the billions and billions of dollars spent annually in the production and improvement of the means of destruction. The highest item in the budget of any nation is the acquisition and development of weapons. The destruction of humans by humans has acquired an aureole of dignity and honor that has replaced the guilt and shame that such an act would imply. The more decorations we exhibit on our chest in recognition of the highest efficacy in smashing and disemboweling people, the happier and prouder we are. The same pleasurable feeling that the head hunter felt when looking at his collection of heads gathered during his incursions in the territory of others is the feeling we experience at the contemplation of the results of a bombing raid on an enemy territory. The more people we smash, the better and more proud we are of our accomplishment. And that is what we call "humanity"? Is this what Darwin calls "the most noble of all attributes of man"? Please. Give me a break!

Human antagonism has been controlled through laws that try to prevent the destruction of the individual and the group. This antagonism is without control, however, when it is manifested at a national level during wartime, and patriotic values are used by governments to achieve their means. These patriotic values usually mask the economic motives of war (page 54).20

All other types and variations of human aggression, including aggression that appears within the densely populated urban centers, such as vandalism and the formation of gangs, have the same origins that have been developed in this chapter, and are only accommodations to different circumstances (page 54).21 The larger a population, the more acute the violence in it. When shielded by anonymity, it is easier to manifest natural instincts, until then controlled by the imposed rules of the society.

Notes

1. *Man the paradoxical Primate*, Jordi Fuentes, 1985.
2. Ibid.
3. Ibid.
4. Ibid.
5. Ibid.
6. Ibid.
7. Ibid.
8. Ibid.
9. Ibid.
10. A writer for the *Spectator*, March 12, 1871, page 320, comments as follows on this passage: "Mr. Darwin finds himself compelled to reintroduce a new doctrine of the fall of man. He shews that the instincts of the higher animals are far nobler than the habits of savage races of men, and he finds himself, therefore, compelled to re-introduce, - in a form of the substantial orthodoxy of which he appears to be quite unconscious, - and to introduce as a scientific hypothesis the doctrine that man's gain of *knowledge* was the cause of a temporary but long-enduring moral deterioration as indicated by the many foul customs, especially as to marriage, of savages tribes. What does the Jewish tradition of the moral degeneration of man through his snatching at a knowledge forbidden him by his highest instinct assert beyond this?"
11. Man, the Paradoxical Primate, Jordi Fuentes, 1985.
12. Ibid.
13. Ibid.
14. Ibid.
15. Ibid.
16. Ibid.
17. Ibid.
18. Ibid.
19. Dissertation on Ethical Philosophy, 1837, page 231, and others.
20. *Man, the Paradoxical Primate*, Jordi Fuentes, 1985.
21. Ibid.

VII

Darwin's Second Omission: Man's Foot Was Never a Hand

Before beginning with this important subject, it will be useful to remember the changes that occur in living creatures when using some organs more than others, or when one organ is not used at all. Bodybuilders know very well that the constant gymnastic exercises modify the muscles, increasing their size, strength, and their efficacy. Let us refer here to some of Darwin's observations with regard to these changes that happen in a short period of time. In the chapter *"Effects of the Increased Use and Disuse of Parts,"* Darwin explains (page 418): *"It is well known that use strengthens the muscles in the individual, and complete disuse, or the destruction of the proper nerve, weakens them. When the eye is destroyed, the optic nerve often becomes atrophied. When an artery is tied, the lateral channels increase not only in diameter, but in the thickness and strength of their coats. When one kidney ceases to act from disease, the other increases in size, and does double work. Bones increase not only in thickness, from carrying a greater weight.* 1 *Different occupations, habitually followed, lead to changed proportions in various parts of the body. Thus it was ascertained by the United States Commission* 2 *that the legs of the sailors employed in the late war were longer by 0.217 of an inch than those of the soldiers, though the sailors were on average shorter men; whilst their arms were shorter by 1.09 of an inch, and therefore, out of proportion, shorter in relation to their lesser height. This shortness of the arms is apparently due to their greater use, and is an unexpected result: but sailors chiefly use their arms in pulling, and not in supporting weights. With sailors, the girth of the neck and the depth of the instep are greater, whilst the circumference of the chest, waist, and hips is less than in soldiers. Whether the several foregoing modifications would become hereditary, if the same habits of life were followed during many generations, is not known, but it is probable. Rengger* 3 *attributes the thin legs and thick arms of the Payaguas Indians to successive generations having passed their whole lives in canoes, with their lower extremities motionless. Other writers have come to similar conclusions in analogous cases. According to Cranz,* 4 *who lived for a long time with the Esquimaux, "the natives believe that ingenuity and dexterity in sea-catching (their highest art and virtue) is hereditary; there is really something in it, for the son of a celebrated seal-catcher will distinguish himself, though the loss of his father in childhood." But in this case it is mental aptitude, quite as much as bodily structure, which appears to be inherited. It is asserted that the hands of English labourers are at birth larger than those of the gentry.* 5 *From the correlation which exists, at least in some cases* 6 *between the development of the extremities and of the jaws, it is possible that in those classes which do not labour much with their hands and feet, the jaws would be reduced in size from this cause. That they are generally smaller in refined and civilized men than in hard-working men or savages is certain. But with savages, as Mr. Herbert Spencer* 7 *has remarked, the greater use of the jaws in chewing coarse, uncooked food would act in a direct manner on the masticatory muscles, and on the bones to which they are attached. In infants, long before birth, the skin on the soles of the feet is thicker than on any other part of the body.* 8 *iIt can hardly be doubted that this is due to the inherited effects of pressure during a long series of generations."* On page 420, Darwin said: *"The Quechua Indians inhabit the lofty plateaux of Peru;*

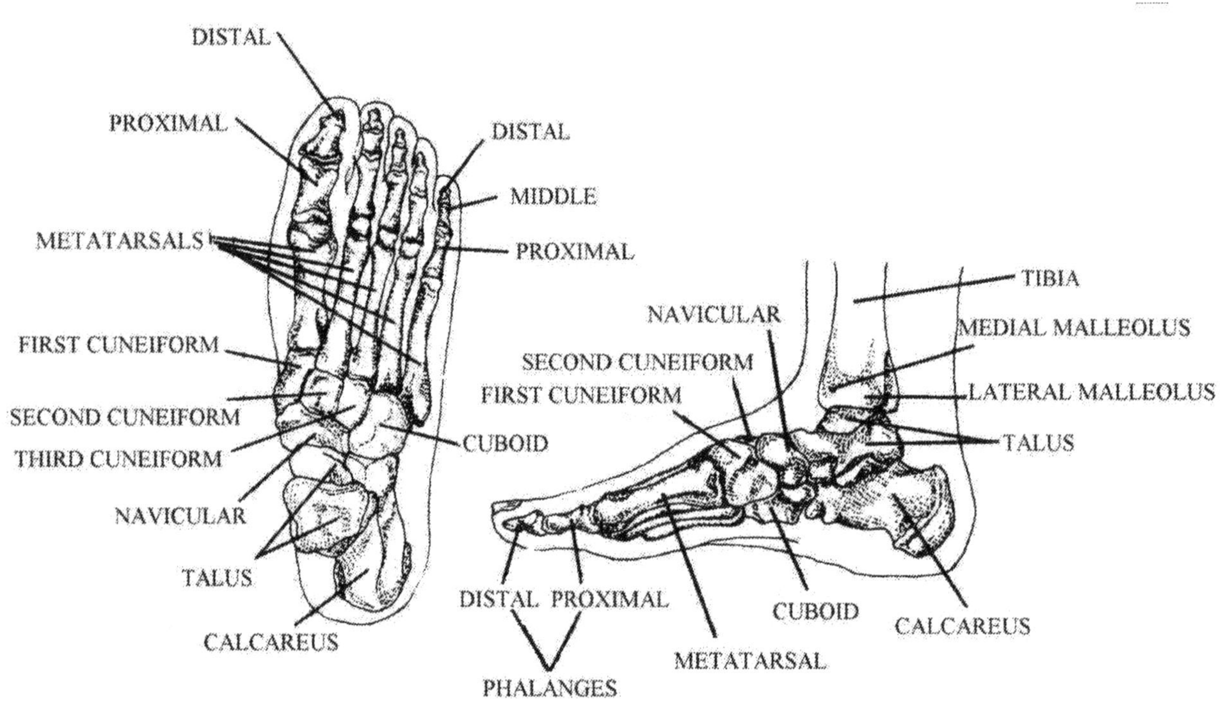

Human right foot. Top view

Right foot. Medial view.

and Alcide d'Orbigny 9 states that, from continually breathing a highly rarefied atmosphere, they have acquired chests and lungs of extraordinary dimensions. The cells, also, of the lungs are larger and more numerous than in Europeans. These observations have been doubted, but Mr. D. Forbes carefully measured many Aymaras, an allied race, living at the height of between 10,000 and 15,000 feet; and he informs me that they differ conspicuously from men of all races seen by him, in the circumference and length of their bodies." 10

With this exposition, Darwin reaffirms Lamarck's theory that the function creates the organ. Nature produces the characteristics needed for the survival of the species, and eliminates those that do not benefit, or that have no meaning or function for the survival of the individuals.

Now let us move on to the explanation for the title of this chapter. At first glance, one can appreciate significant differences between human and primate feet. The structure and the function of their feet are distinct (page38). But Darwin said (page 395): *"It is notorious that man is constructed on the same general type or model as other animals. All the same bones in his skeleton can be compared with corresponding bones in a monkey, bat, or seal. So it is with his muscles, nerves, blood vessels, and internal viscera. The brain, the most important of all organs, follows the same law, as shewn by Huxley and other anatomists. Bischoff,* 12*), who is a hostile witness, admits that every chief fissure and fold in the brain of man has its analogy in that of the orang; but he adds that at no period of development do their brains perfectly agree; not could perfect agreement be expected, for otherwise their mental powers would have been the same."*

I agree that the general structure is the same in vertebrate animals and man, but this general pattern has been adapted to many different styles of life. We can appreciate the general structure, but we also distinguish the differences held

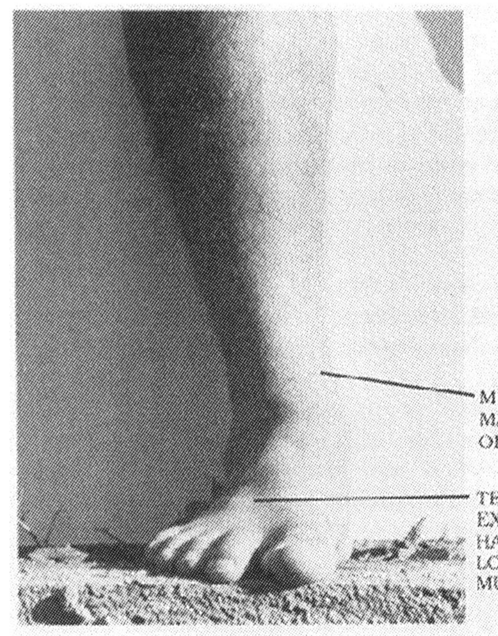

GASTROCNEMIUS MUSCLE

SOLEUS MUSCLE

MEDIAL MALLEOLUS OF TIBIA

CALCANEAL TENDON

LATERAL MALLEOLUS OF FIBIA

Muscles of human leg and ankle
Surface anatomy photograph

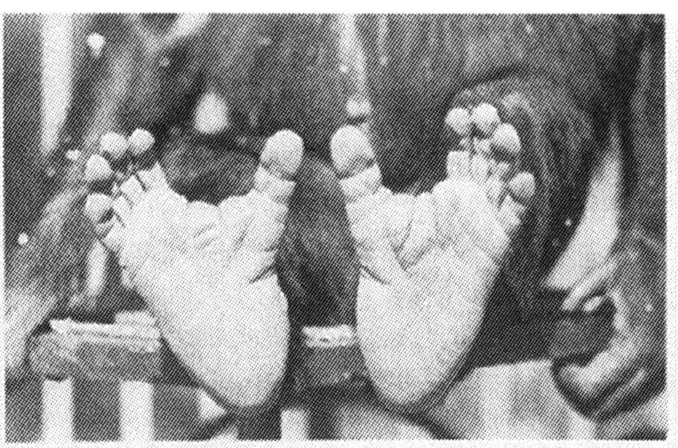

a

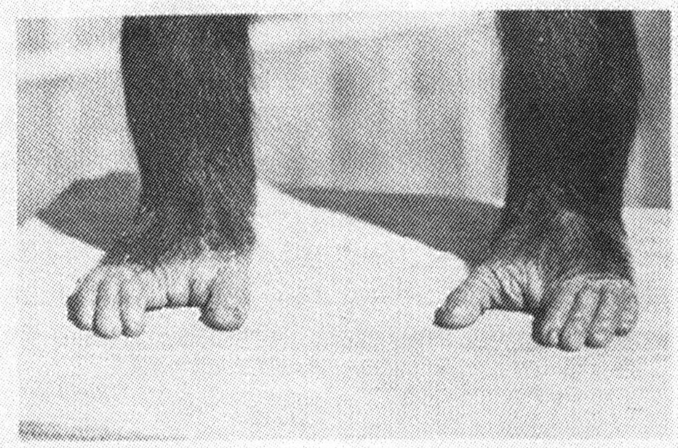

b

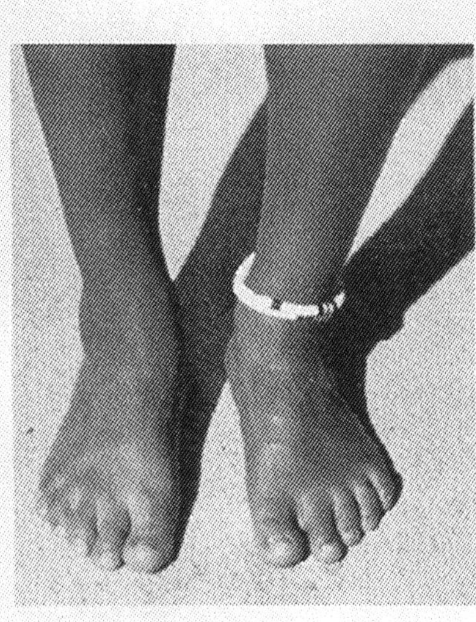

c

Toe placement
a) Bottom view of chimpanzee feet. The great toes (hallux) are well defined and functional allowing the chimpanzee to use his feet as hands. b) Top view of chimpanzee feet. The great toe is completely separated from the other toes. c) top view of human feet.

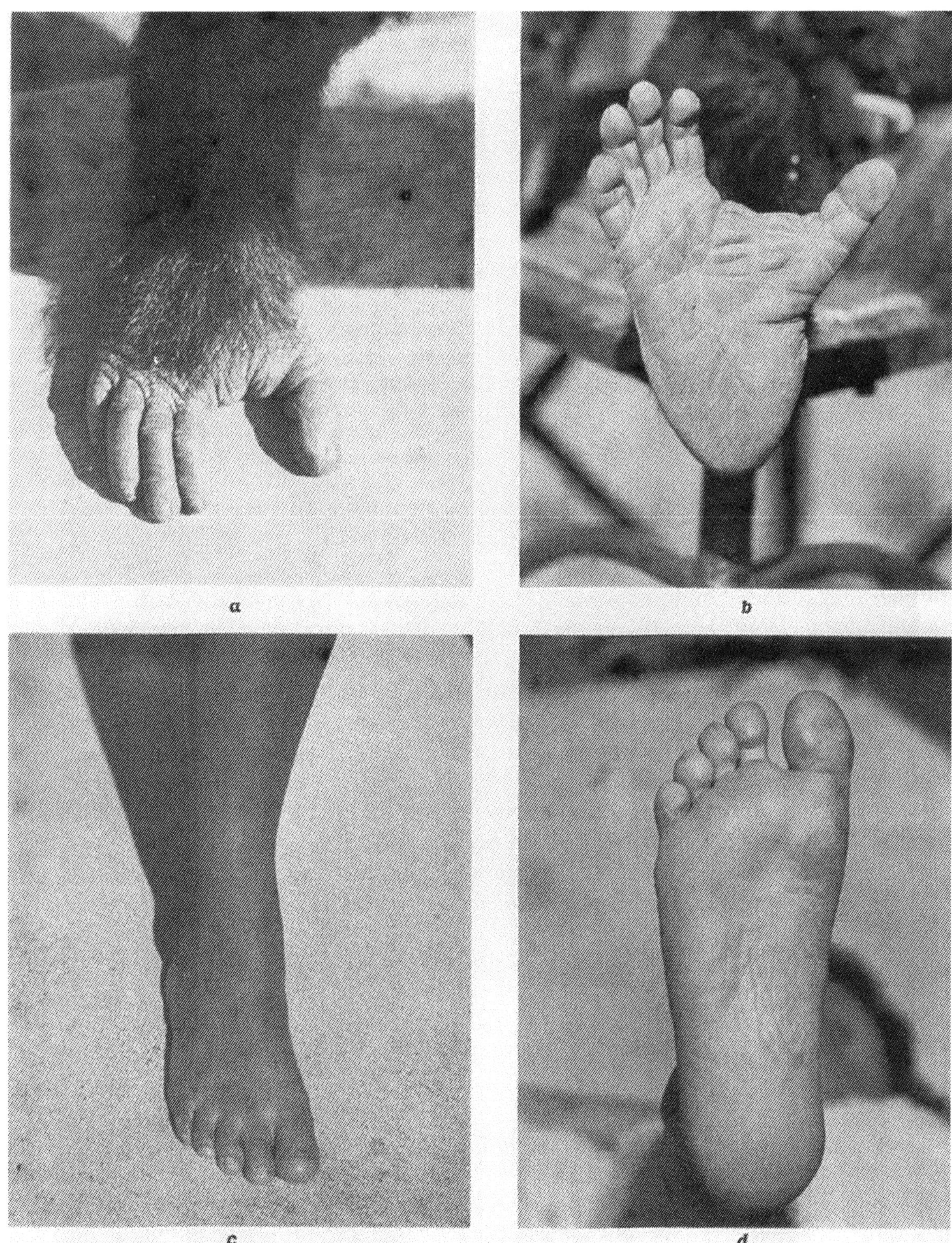

Toe placement

a) Bottom view of chimpanzee feet. The great toes (hallux) are well defined and functional allowing the chimpanzee to use his feet as hands. b) Top view of chimpanzee feet. The great toe is completely separated from the other toes. c) top view of human feet today as adaptations to the different ecological niches in which each animal species lives. The idea of a general pattern between animals and man was so fixed in Darwin's mind, that it prevented him from paying enough attention to the enormous differences that we can appreciate in the evolution of analogous organs between one species and another, using his own theory of evolution. He missed important characteristics that show that similar organs underwent different evolutionary patterns according to the functions expected from them in the adaptation of the animal to his survival needs. One of the organs where he missed the different evolution from one organ to the production of two very different structures is the anthropoids foot and the human foot.

The anthropoid's foot is really a hand, and includes a well-defined thumb. The human foot is not a hand in any sense of the word and does not possess an opposable thumb, but rather an enlarged toe (hallux) that otherwise is exactly like the other four toes on our foot. The big toe is larger because it offers more support and must sustain the entire weight of an individual upon walking; the heel is lifted from the floor and body weight is passed from the arch, through the joints of the toes to the big toe. At this point, all the body's weight rests on one toe, as the other toes are not offering surface support (page38). 13

Toe articulation is necessary to be able to walk. The sole of the foot is rigid, as it is the surface that allows the body to be sustained in an upright position. The quadrupeds do not need a sole, and the extremities of their feet have been reduced to such a degree that some animals, in fact, walk using mainly the tips of their nails or claws. We may observe a similar illustration of weight distribution with a table. No matter how delicate or thin the legs of a four-legged table are, it will always stand up. If, on the other hand, the table only has two legs, these legs necessarily need an extension to distribute the weight so the table does not fall. In other words, this table needs a "sole" to stabilize weight distribution. If the sole is rigid, it is hard to rotate the foot from the ground. This is the difficulty encountered by individuals who, for one reason or another, have lost their toes. To be able to walk easily, it is necessary to have a point of articulation with which the body can separate itself from the weight-sustaining surface (floor). This articulation is found between each metatarsus and the first phalanx. The weight is passed from one toe to the other until arriving at the big toe, where the final separation from the floor occurs. This movement is what makes it necessary for the toes to be short; if the toes were longer, more articulations would be necessary to carry out the same function (page 38 and 39). 14

The human foot does not exhibit specialization for arboreal life, and the human foot is not prehensil (a hand). Further, it would seem to me that:

If the human foot were, in fact, a derivative of a hand, our foot would only have four toes, and not five.

Why would I suggest this? Here is why.

In human hands, the thumb or pollex, is located closer to the wrist than are the rest of the fingers, more or less within the first third of the palm. The same happens with the other anthropoids. If this hand had been used to walk in a bipedal position, the thumb would not have touched the floor. The thumb would have been separated from the other fingers and would no longer support weight. Upon losing its function, it would very likely disappear. The other four fingers would be those to sustain the body's weight upon separating the hand from the floor, with the emphasis on the extreme articulation. At this point, the fingers would no longer be fingers, but toes (page 39). 15

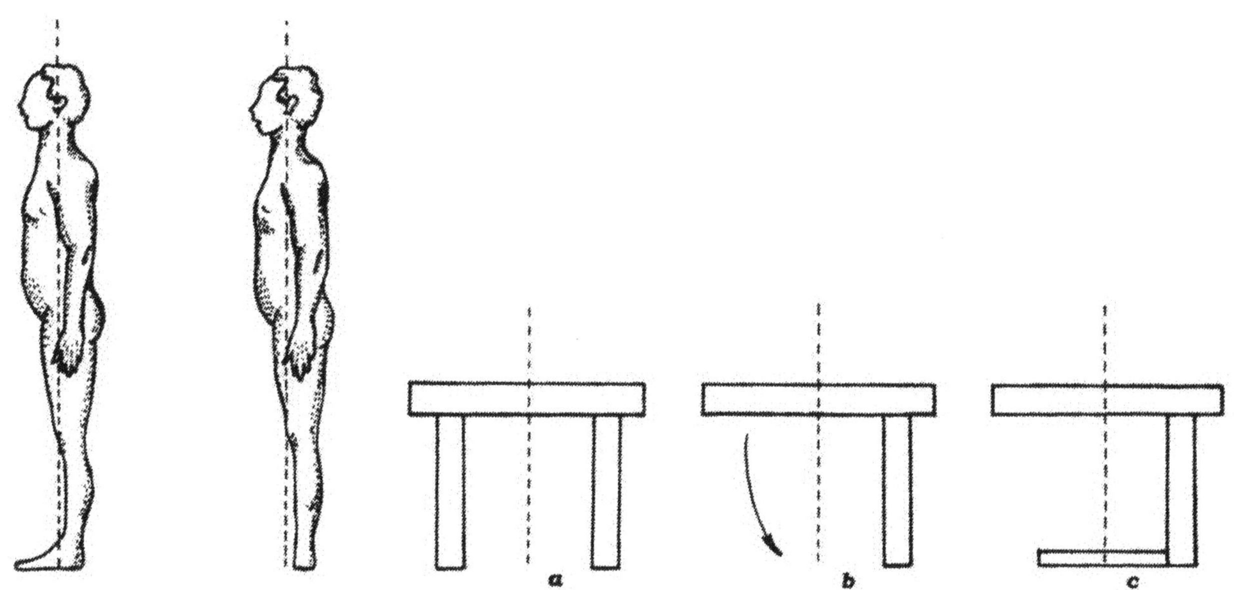

1. Upright position of the human body with "human" foot. Center of gravity in the support (sole). Stable. Case similar to (c). 2. Upright position of the human body with soleless foot. Unstable. Center of gravity out of support. Impossible balance. Case similar to (b). (a) A four-legged table stands up. Center of gravity between supports. Stable. (b) A two-legged table. Center of gravity out of support. Unstable. (c) A two-legged table with a "sole". Center of gravity in the support. Stable.

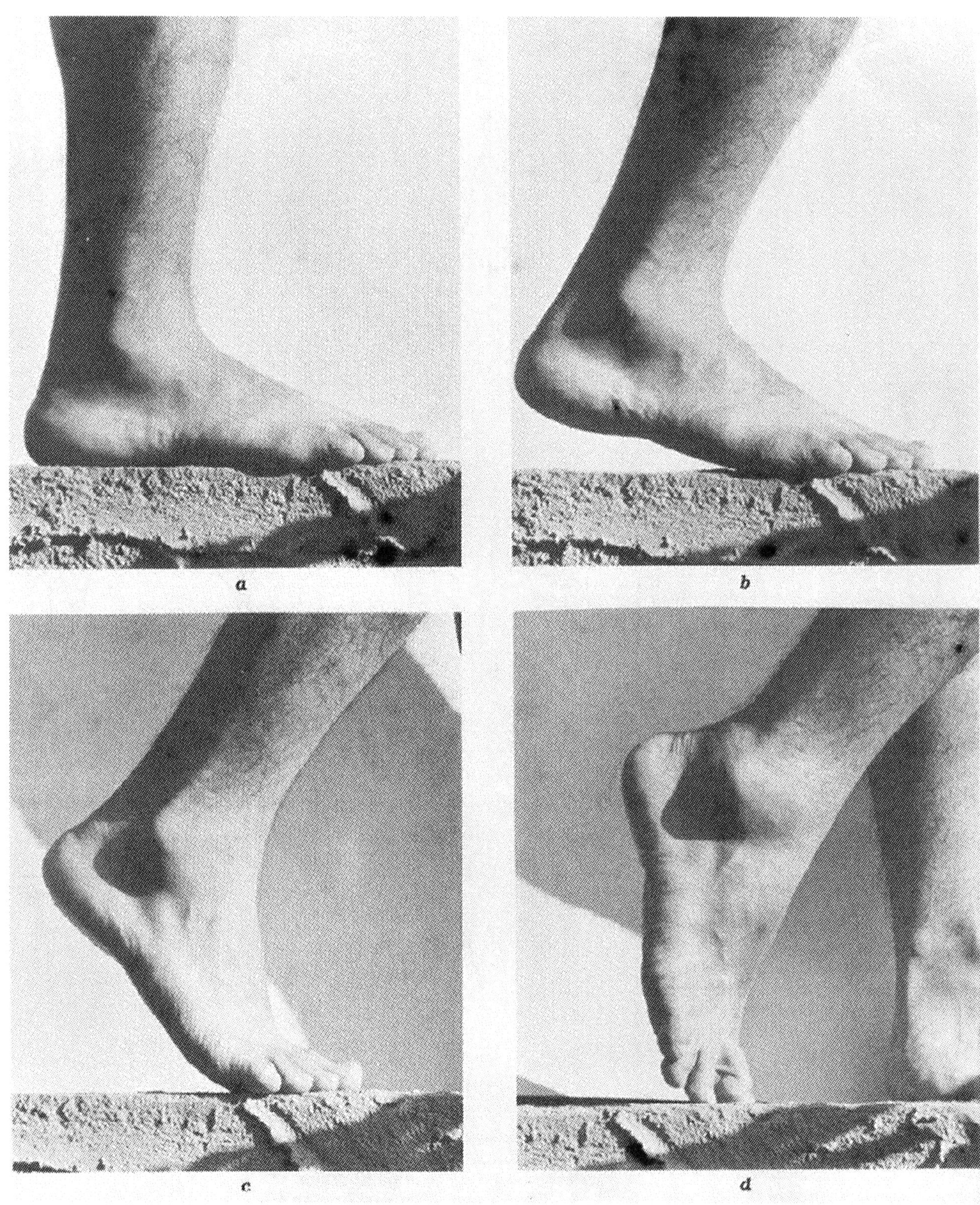

When walking the heel is lifted from the floor and the body weight is passed from the arch, through the joints of the toes to the large toe (Hallux). At this point all the body's weight rests on one toe and the other toes are not offering support. a) Foot at rest. Weight evenly distributed. b) Initial forward movement. Weight transferred to the toes. c) Weight passes from small toes . . . to large toe d).

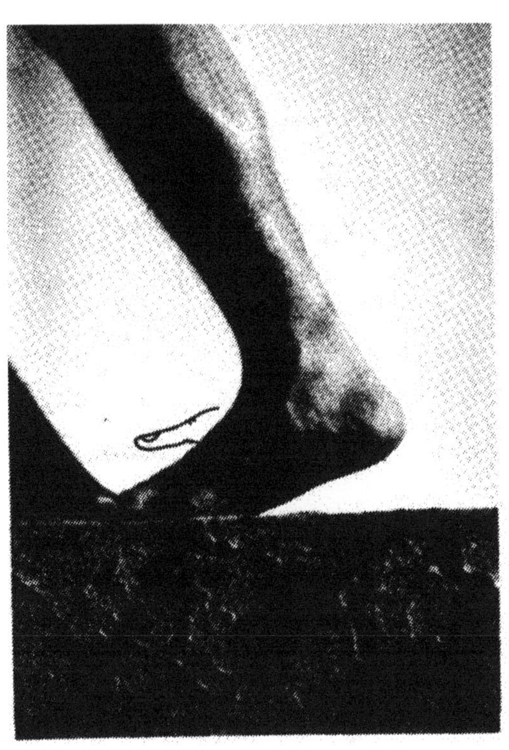

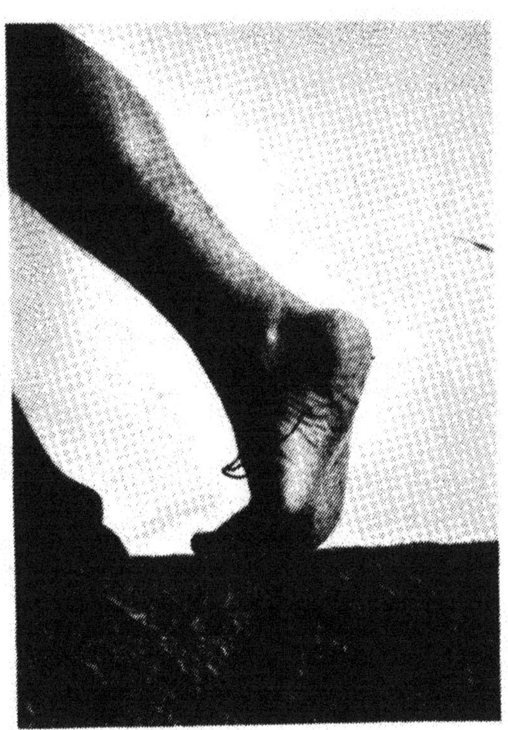

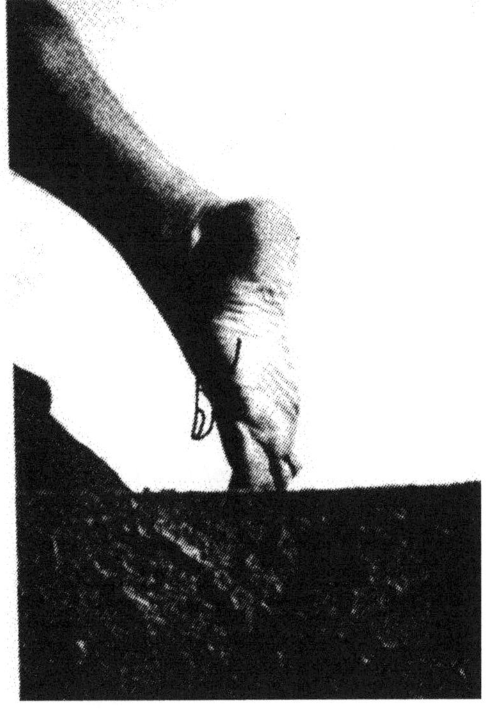

Position of the thumb relative to the floor

If the foot had been a derivative of a hand, the thumb would not support weight, and upon losing its function, it would very likely disappear and the foot would have only four toes instead of five.

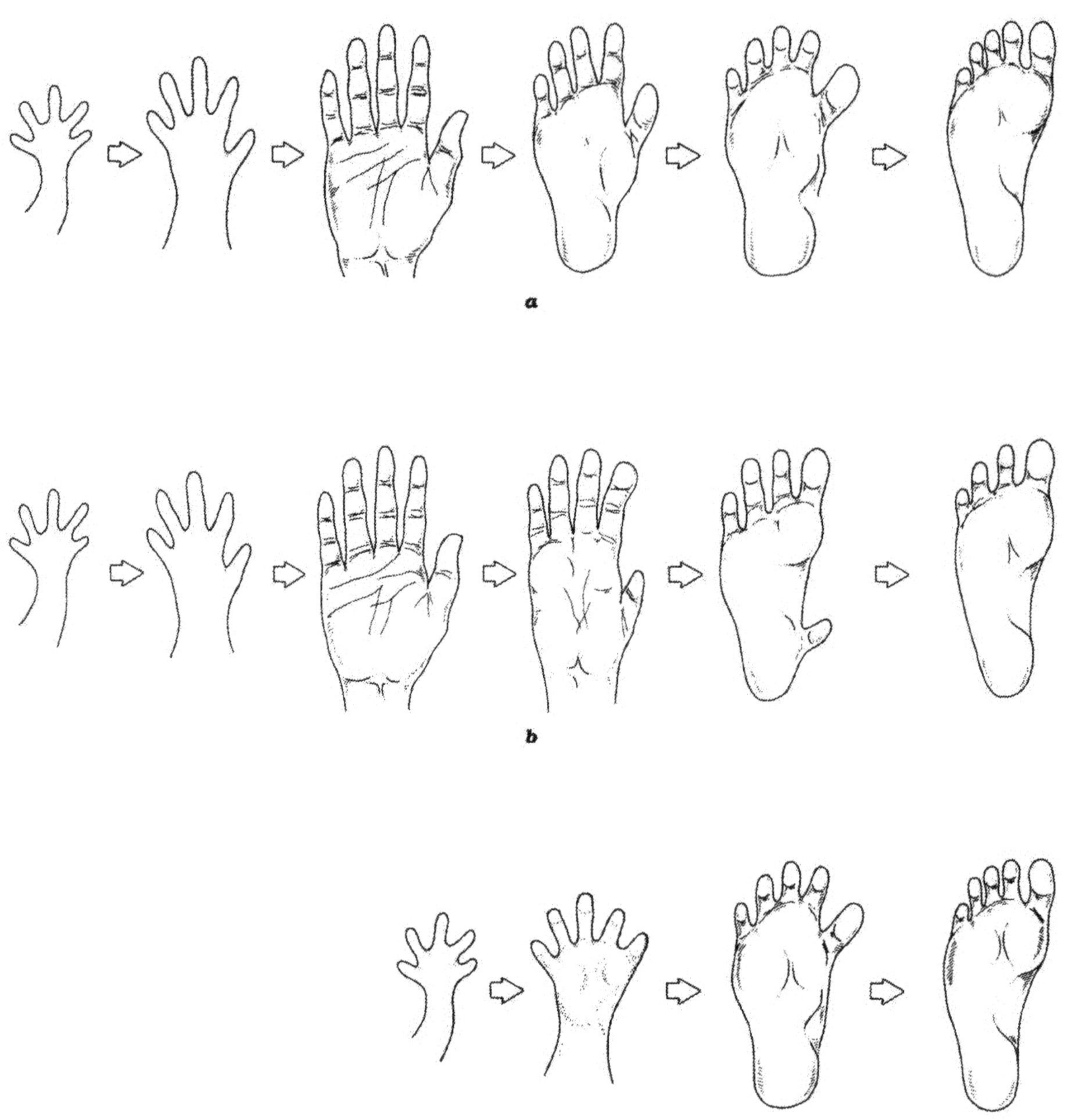

Formation sequences

a) Formation sequence of the humanfoot from the amphibian foot according to the most prevalent theory today. From the amphibian foot the hand would have evolved (four hands for arboreal life) and from the posterior hands the foot would have originated when man's ancestor evolved towards bipedalism.

b) Formation sequence of the humanfoot from the amphibian foot. If the human arboreal ancestor developed two feet from four hands – then the resulting feet should have only four toes.

c) Formation sequence of the human foot from the amphibian foot. This sequence would leave out the formation of the hand, the foot developing gradually from the original amphibian form to the present human foot conserving the five original digits. The human hand and human foot would have developed simultaneously but separately from the amphibian foot to the present forms of specialization.

Other animals, including primates, have lost their "thumbs" or had them. In the case of the human foot, if it had evolved from a hand, the inside toe of the four toes left would be the one to enlarge in order to support the body's weight in the act of walking (page39).16

Darwin states, in what he calls "Correlated Variation" (page 427), *"In man, as in the lower animals, many structures are so intimately related that when one part varies so does another, without our being able, in most cases, to assign any reason. We cannot say whether the one part governs the other, or whether both are governed by some earlier developed part.."*

When Darwin was writing that paragraph, he accepted that he did not know the reasons for this correlation between different parts or organs. He was too advanced for his time. When he was wondering about the possible connections between different organs in the body, the Austrian monk Gregory Mendel was experimenting with plants in the Augustinian monastery of Brünn. His experiments proved that there is a definite pattern in the way contrasting characteristics are inherited. He published the results of his experiments in 1866. The important laws of heredity, were clearly exposed in his writing. Unfortunately, Darwin did not know about Mendel's work. These discoveries laid the foundation for the scientific study of heredity, but unfortunately Mendel's work was not discovered until the beginning of the twentieth century. Darwin suspected the existence of this mechanism but he did not know it. As a consequence, he did not have the opportunity to learn about the existence of chromosomes, or that genes grouped in DNA (deoxyribonucleic acid) inside the chromosomes. In the middle of the twentieth century (1957), the works of Francis H. C. Crick and James Watson produced a model showing the structure of DNA. The genetic code was not discovered until 1966. Genes are tiny particles that determine all hereditary traits. Those genes have a specific function in determining the different characteristics of the individual. At the same time, one gene controls or influences more than one characteristic. It influences other characteristics carried in other genes. Unfortunately, all these discoveries came one century later. They could not be used by Darwin, so he could not explain the points that he sensed were not exhaustively explained.

Natural selection acts in only one direction in the sense that once something is initiated it is not retracted.

The amphibians are considered the first animals to actually walk on land. In order to do this, they developed four feet with five toes each. Why five toes and not another number? Nobody knows. Presently, amphibians continue, generally, to have five toes.

If, from this five-toed foot without differentiation between the digits, nature wanted to create a hand, natural selection would have to displace the innermost digit (toe) and move it outward toward the wrist within the first third of the palm away from the other digits and opposable to them. This natural selection would be obeying a necessity to create an extremity whose function would be distinct from walking. If the human foot had been derived from a hand just described, nature would have had to undo all the changes it had accomplished and move the toe or thumb back toward the other digits. Why would nature displace this toe or thumb if this did not create any advantage for the animal? It is much easier to suppress this digit. Nature has used this solution frequently, and an extreme example of this is illustrated by the perissodactyl herbivores represented by the horse, where all of the toes have become atrophied save one (pages 39 and 41).17

Here I will repeat the assertion stated before: Natural selection acts only in one direction. Natural selection is not playing tricks; once something is done, it is done. If the finger was carried away from the other four fingers in order to create a hand, Natural selection will not move it backward and undo all this process to assume the initial position again. The intermediate stages between front and back would have no vital reason for the species, so natural selection would not favor this return to the initial position.

Following this line of thinking it can be deduced that:
If the human foot was derived from a hand, it would only have four toes. As it has five toes, this is a suggestion that the origin of the human foot was not a hand, but rather and amphibian-like foot.

The atrophy or loss of digits is well represented, as mentioned previously, among the primates. The genus *Ateles*, two species known as the spider monkeys are so specialized in arboreal life that two characteristics have developed: the prehensile tail and their modified hands, which are a kind of "hook" with elongated palms, and curved, long fingers that lack thumbs (pollices) (page 41).18 With respect to the atrophy of organs, Darwin says (page 400): *"Not one of the higher animals can be named which does not bear some part in a rudimentary condition; and man forms no exception to the rule. Rudimentary organs must be distinguished from those that are nascent; though in some cases the distinction is not easy. Nascent organs, on the other hand, though not fully developed, are of high service to their possessors, and are capable of further development. Rudimentary organs are eminently variable; and this is partly intelligible, as they are useless, or nearly useless and consequently are not longer subject to natural selection."*

Here, Darwin made a mistake. He contradicted himself because he stated that natural selection accentuates those organs useful for the survival of the species, and eliminates those that have no meaning in the struggle for survival. If he says that natural selection eliminates those rudimentary organs, *"as they are useless,"* how can they be, at the same time, *"no longer subject to natural selection"*? A good example of this elimination in the primates is his following comment (page 433): *"for the most arboreal monkeys in the world, namely, Ateles in America, Colobus in Africa, and Hylobates in Asia, are either thumbless, or their toes are partially cohere, so that their limbs are converted into mere grasping hooks."* According to Darwin, an organ that becomes useless is eliminated by natural selection, in all animals; in the case of primates, with the thumbless *Ateles*, the thumb became a nuisance in its movements in the canopy of the trees. Then if an organ is eliminated because it is useless, how it can be *"no longer subject to natural selection?"* He did not reflect about this point. In reference to the reduction and further elimination of organs, Darwin says (page 400): *"The chief agents in causing organs to become rudimentary seems to have been the disuse at that period of life when the organ is chiefly used (and this is generally during maturity) and also inheritance at a corresponding period of life. The term "disuse" does not relate merely to the lessened action of muscles, but includes a diminished flow of blood to a part or organ, from being subjected to fewer alterations of pressure, or from becoming in any way less habitually active. Rudiments, however, may occur in one sex of those parts which are normally present in the other sex, and such rudiments, as we shall see, have often originated in a way distinct from those referred to. In some cases, organs have been reduced by means of natural selection, from having become injurious to the species under changed habits of life. The process of reduction is probably often aided through the two principles of compensation and economy of growth; but the later stages of reduction, after disuse has done all that can fairly be attributed to it, and when the saving to be effected by the economy of growth would be very small, are difficult to understand. The final and complete suppression of a part, already useless and much reduced in size, in which case neither compensation nor economy can come into play, is perhaps intelligible by the aid of the hypothesis of pangenesis.* 19 [The reader is referred to the definition of "pangenesis" in the Glossary] *But as the whole subject of rudimentary organs has been discussed and illustrated in my former works, 20 I need here say no more in this head."*

The conclusion I arrived at after the study on the functionality of the foot is that:

The human foot is an instrument adapted exclusively for terrestrial locomotion. The human is, and always has been, a walking animal.

Notes

1. I have given authorities for these several statements in my *Variation of Animals under Domestication.* vol. ii, pp. 297-300. Dr. Jaeger, *Ueber das Längenwaschsthum der Knochen, Jenaischen Zeitschrift,* B.v. Heft i.
2. *Investigations*, B. A. Gould, 1869, p 288.
3. *Säugethiere von Paraguay*, 1830, s. 4.
4. *History of Greenland*, English translation 1767, vol. i p. 230.
5. *Intermarriage.*, Alex. Walker, 1838, p. 377.
6. *The Variation of Animals under Domestication*, vol. i, p. 173.
7. *Principles of Biology*, vol. i p. 455.
8. Paget, Lectures on Surgical Pathology, vol. ii, 1853, p. 209.
9. Quoted by Prichard, *Researches into the Physical History of Mankind*, vol. v. p. 463.
10. Mr. Forbes' valuable paper is now published in the *Journal of the Ethnological Society of London.* New series. Vol. ii 1870, p. 103.
11. *Man, the Paradoxical Primate*, Jordi Fuentes 1985.
12. *Grosshirnwindungen des Menschen*, 1868, s. 96
13. *Man, the Paradoxical Primate*, Jordi Fuentes 1985.
14. Ibid.
15. Ibid.
16. Ibid.
17. Ibid.
18. Ibid.
19. *Brosshirnwindungen des Menschen*, 1868, s. 96.
20. *Variation of Animals and Plants under Domesticiation* vol. ii. p. 317. See also *The Origin of Species*, p. 346.

VIII

Man, a Case of Neoteny

Man is one of the few animals that exhibits neoteny. **Neoteny is the retention of juvenile or immature characteristics through adulthood.** Adulthood here is understood as a fully developed and mature state in which sexual maturity has been reached.

In most animals, sexual maturity is not reached until the animal is fully grown and developed. Increased weight experienced by many mammalian animals after sexual maturity is not true growth, but the addition of adipose tissues. A good example of neoteny may be seen in the Mexican amphibian axolotl (*Ambystoma mexicanum*), which reproduces itself while still exhibiting the external appearance of the larval state. This organism does not experience metamorphosis characteristic of many amphibians, during which the gills are lost and lungs are developed; the tail is reduced, the neck becomes distinguishable, adult coloring is taken on, and there is the ability to live on land. The axolotl apparently reproduces itself while still in a larval state. The word "apparently" is used here because what, in fact, has happened is that the animal has reached maturity, but in this case, the lack of thyroxine secreted by the thyroid gland that controls metabolism prevents the transformation of the organism (page 19).1

Man's capability to reproduce himself appears long before physical development has been concluded (page 19).2 This is another point of discrepancy between Darwin's statements, and my humble opinion. Darwin says (page 398): *"It has been urged by some writers, as an important condition, that with man the young arrive at maturity at a much later age than with any other animal: But if we look to the races of mankind which inhabit tropical countries the difference is not great, for the orang is believed not to be an adult till the age of from ten to fifteen years."*

In reference to this point, Darwin understands the arrival to maturity in absolute numbers, not in relation or proportion to the life span of the animal under discussion. He interprets as "maturity" the arrival at puberty or sexual activity that goes together with the physical maturity of the referred animal. Darwin does not differentiate puberty from physical maturity. If we calculate using absolute years without considering the life span of the animal, such as Darwin did, there are many cases where humans reach "maturity" before other animals. Human females reach puberty and are able to bear children, in some cases, as early as ten to twelve years old. Crocodiles don't' reach sexual activity until fifteen years old, though they have already reached adult size, but reptiles continue to grow all their lives. The desert California tortoise (*Gopherus agassizi*) is not able to reproduce before it is fifteen.

Darwin ignored the existence of neoteny in the human being. I will insist on this point: I will call puberty "sexual readiness," not body maturity.

Puberty is reached in the human beings as early as ten years for girls, and twelve for boys. There are exceptions in both cases; biology is not an exact science, and adaptations and exceptions abound. These ages are an average. Complete physical development and growth is not reached in man until the mid-twenties. Anthropoids do not reach puberty until they reach complete physical growth. Gorilla males are sexually mature at about eight years, and females at about six. Their record life span in captivity is held by "Bamboo," a male gorilla at the Philadelphia Zoo that lived to be thirty-eight. Orangutans reach puberty at about ten years, and their life span is about thirty-five. The increase of weight shown by male gorillas and orangutans after reaching maturity is only an accumulation of fat, but not an increase in size.

Once this point is established, we are left wondering why man is such an exception among anthropoids. The answer must be that neoteny has offered man advantages for survival. But which ones?

Let's analyze the situation. Man undergoes very slow development both in the fetal stage and after birth. After birth, he is still in a relatively helpless stage of development. Because of this retarded state of development, man is obligated to undergo a prolonged state of infancy until he learns to use all parts of his body, and to use and develop his mind and language. Mental maturity develops slowly in man. A teenager has a very developed body but does not have mental maturity, at least until he or she is twenty-four (remember that there are exceptions to the rule). This is the reason that insurance companies have special rates for teenagers, until they show proficiency in their driving.

It is fairly safe to assume that *Homo sapiens'* life expectancy used to be short. Animals that live in the wild have life spans that are much shorter than the same animals kept under domestic conditions. In the wild, they are subjected to predators, diseases, scarcity of food, natural disasters, and intemperate climatic conditions. All of these adverse conditions, of course, shorten life expectancy. Our ancestors had to face predators and conditions that today, for the most part no longer exist. Human life expectancy has only recently prolonged itself with the massive production of agricultural products, the ability to fight diseases, and, even more important, the ability to prevent disease.

The most impressive achievements in longevity have taken place in this century more specifically, in the last few decades.

Most *Homo sapiens* fossils and skeletons known today were of individuals who died relatively young. The hominid fossils also reveal this peculiarity. *Lucy*, the most complete skeleton of any erect walking human ancestor, found by Dr. Donald Johanson in Afar, Ethiopia, is estimated to have been approximately twenty-five years old at death. Was this a young age? Surely not in relative terms, and it is very possible that twenty-five was the average life expectancy 3.5 million years ago for this species (page 19).3

Here is where neoteny is important. If man had to achieve full physical maturity and growth before reproducing his kind, surely his kind would have disappeared long ago, for the lack of time to raise the children. There were too many threats to his life, and he simply wouldn't have had enough time to reach mature physical growth and raise children to a point where they were self-sufficient after a long infancy. Neoteny allows man to reproduce himself and raise his offspring while he continues to grow within his family group, and his evolutionary characteristics favor his survival as a species (page 20).4

Seen from this point of view, neoteny could have evolved millions of years ago within an evolutionary branch predating hominids, and could have passed on to hominids, and from hominids to *Homo sapiens sapiens* (page 20).6 This ability to reproduce oneself while still an adolescent has created many headaches for modern-day parents; early and undesirable pregnancies are common in our Western culture.

Let us consider this situation: The fact that man's offspring is born in a little-developed state and the existence of neoteny are two facts closely related. If, once, there was a branch of *Homo*, or of an australopithecine or earlier ancestor who did not manifest neoteny, then it is possible that this branch would have died out. Lateral branches lacking neoteny but having the ability too develop rapidly did have the ability to survive and most likely gave rise to the simians. Anthropoids whose offspring were born in an advanced state prevented them from evolving into the true *Homo*.

Neoteny and Tthe Human Head

It has been suggested that man exhibits an example of neoteny because he conserves the infant-like roundness of his head in adulthood. As supporting evidence for this hypothesis, it has been pointed out that there is much more resemblance between human and chimpanzee heads during the fetal stage. The adult chimpanzee head has undergone a greater change in shape from the fetal stage than has its human counterpart. The adult chimpanzee head is much more elongated. In the case of humans, there is really very little difference between the shape of the head in the fetal or the adult stage.

Another point in opposition to this hypothesis is that it is a general rule in biology that all mammalian infants have a tendency to exhibit a roundness in all their body parts, which is lost as the body develops. This tendency is also found in nidifugous birds and, to a lesser degree, in some reptiles.

What I suggest here is that the rounded shape of the human head and face is not so much a case of neoteny as it is the product of evolution, making allowance for greater cranial capacity and room for facial muscles necessary for us to speak. (See chapter XI section, Funtional Areas of the Cerebrum, and chapter XII, the Beginning of Language.)

All animals that possess hands and hand-like front paws, used to put food into the mouth, have a tendency to have a more rounded head, as an elongated snout is not necessary. Some examples include squirrels, chipmunks, prairie dogs, pakas, otters, lemurs, and monkeys. The roundness of the head does not seem to be a consequence of neoteny.

Notes

1. *Man, the Paradioxical Primate*, Jordi Fuentes, 1985.
2. Ibid.
3. Huxley, *Man's Place in Nature*, 1863, p. 34.
4. *Man, the Paradioxical Primate*, Jordi Fuentes, 1985.
5. Ibid.
6. Ibid.

IX

Hairless Man

Man does not posses as much hair as do most other mammals, where their hair covers most of their body. There are mammals that have little hair, such as the aardvark (*Orycteropus afe*r), which is sparsely covered with hair; the hippopotamus (*Hippopotamus amphibius*), which is hairless except for sparse bristles on the tail, ears, and muzzle; and certain African underground rodents that only manifest vestiges of hair on their snouts. Other than these animals, only cetaceans and the order *Sirenia* may be considered hairless. Only a few hairs around the snouts of adult baleen whales and toothed whale (*Physeter catodon*) fetuses are present. The loss of hair in these animals is not well understood. It has been suggested that in the case of the African underground rodents, the possession of hair was a disadvantage to their life underground. Why could it be a disadvantage when there are many other underground rodents that have dense fur? We don't have a logical explanation for that. In the cetaceans, hair became unnecessary, perhaps giving way to a more hydrodynamic body. I personally do not necessarily agree with this reasoning in either case, because there are many hydrodynamic marine mammals that possess hair and many underground rodents and insectivores that possess hair (page 21).1

With respect to the hairless condition of the human being Darwin expresses his thinking on the subject in these words (page 406): *"Man differs conspicuously from all the other primates, in being almost naked. But a few short straggling hairs are found over the greater part of the body in man, and fine down on that of a woman. The different races differ much in hairiness; and in the individuals of the same race the hairs are highly variable, not only in abundance, but likewise in position : this in some Europeans the shoulders are quite naked, whilst in others they bear thick turfs of hair.*2 *There can be little doubt that the hairs thus scattered over the body are the rudiments of the uniform hairy coat of the lower animals. This view is rendered all the more probable, as it is known that fine, short, and pale-coloured hairs on the limbs and other parts of the body occasional become developed into "thickset," long, and rather coarse dark hairs", when normally nourished near old-standing inflamed surfaces."* 3

Rather than explain this disappearance of hair through the Lamarckian theory, which is that environmental changes cause structural changes in animals and plants and these changes are passed on to the offspring, or, as the saying goes, "Function creates the organ," this hair was most likely eliminated by natural selection in those animals that are notorious for the lack of it. There can be other reasons for that. In any event, whatever the cause, the result is that these mentioned animals have very little hair. The presence of hair on modern-day man in the armpits, on the head, and in the pubic region has been explained by saying that "these are remnants of hair from a time in the past when hominids' entire bodies were covered in hair." I doubt this was the case (page 21). 4 First, there is no reason to believe that at one time man or his ancestors were abundantly covered with hair and then lost it. Second, why would the remnants of hair persist in those areas where hair is least needed? These should be the areas where hair would be lost first. Many theories have been developed to try to explain this loss of hair. One theory holds that since cetaceans have no hair, perhaps man's ancestors were aquatic. Our ancestors were very likely hairless, otherwise, the offspring would have probably taken advantage of this treat, clinging to his mother's hair, and the little-developed newborn of present man would not exist.

With regard to the lack of hair, Darwin says (page 438): *"Another most conspicuous difference between man and the lower animals is the nakedness of the skin. Whales and porpoises (Cetacea), dugongs (Sirenia), and the hippopotamus are naked; and this may be advantageous to them for gliding through the water; nor would it be injurious from the loss of warmth, as the species, which inhabit the colder regions, are protected by a thick layer of blubber, serving the same purpose as the fur of seals and otters. Elephants and rhinoceroses are almost hairless; and as certain extinct species, which formerly lived under the Arctic climate, were covered with long wool or hair, it would almost appear as if the existing species of both genera have lost their hairy covering from exposure to heat. This appears the more probable, as the elephants in India which live on elevated and cool districts are more hairy than those on the lowlands. May we then infer that man became divested of hair from having aboriginally inhabited some tropical land? That the hair is chiefly retained in the male sex on the chest and face, and in both sexes at the junction of all four limbs with the trunk, favours this inference on the assumption that the hair was lost before man became erect; for the parts which now retain hair would then have been most protected from the heat of the sun. The crown of the head, however, offers a curious exception, for at all times it must have been one of the most exposed parts, yet it is thickly clothed with hair. The fact, however, that the other members of the order primates, to which man belongs, although inhabiting various hot regions, are well clothed with hair, generally thickest on the upper surface, is opposed to the supposition that man became naked through the action of the sun."*

The most hairy region of the human body is the head, and it is possible that this hair is just protection from the sun. But there are several African races living in the tropics, where sun radiation is the most intense, and they show short hair on their heads, but very curly. Is this curliness a protection to prevent the sun from reaching the scalp? I don't know. It is a curious fact that the African human races with little and short hair on their heads, are usually less hairy also on their faces than the European races.

In the case of the cetaceans, there is no necessity for hair, since they are entirely aquatic animals never coming out of the water unless they are going to die. In cetaceans, hair is not necessary to protect them from cold because they have developed a thick layer of blubber under the skin, which insulates them from the low temperatures of the water. Pinnipeds (seals, walruses, and otaridae), on the other hand, are aquatic mammals with a thick layer of blubber under the skin that spend a lot of time out of the water, and they do have a thick coat of hair. Perhaps their coat of hair is protection not against the cold but against the sun. this coat of hair has been the perdition of some species of seals, as they have been hunted almost to extinction for their fur. Walruses are the pinnipeds that have the least amount of hair, perhaps, because they live in northern areas where the sun's rays are oblique to the Earth's surface and are consequently not extremely strong. But what about Arctic seals? They also live in the most northern regions and have thick fur. There is no explanation, so far, to satisfy our interest or curiosity.

Returning to the main point, we can be almost certain that hominids did not have abundant hair. The hominids walked in an upright position as we do, and for the same reasons. They had to hold their offspring in their arms, and their offspring did not have prehensile feet nor sufficient strength to be able to hold on to their mothers. Hair on the chest region and stomach of the mother hominid would not have had a practical function in this respect. Hence, there was no need for it.

One could say at this point that hair on the hominid would not serve to allow the offspring to hold on to his parents, but rather would serve as protection against the elements, such as it does for many animals. However, the irrefutable reality is that man does not need a natural coat to protect himself against the elements, because he uses fur from other animals and has even learned how to weave clothing for this purpose. We know that *Homo* could make his own clothes, but the central question here is whether hominids, his ancestors, had hair or not.

Hair is not, al least for present man, a way to protect himself against the cold. As we know, there is no human society that lives in cold weather that has developed hair for this purpose. According to the theory of Jean Baptiste Lamarck, man's lack of developing hair as a consequence of living in the cold (function) would be because he uses clothes, and therefore the "organ" would not be necessary. This apparently logical explanation has a fault or drawback. We have the example of three tribes of South American Indians, who have lived in the most inhospitable cold [up to twenty-five below zero (centigrade)] and nevertheless have not developed hair. Did they use clothes? No; absolutely no clothes. One of these tribes was the Ona. These Indians lived in the Tierra del Fuego and the Beagle Canal in the extreme south of Patagonia, Argentina. The Ona wore very little clothes except for a guanaco fur around their waist and up over their left shoulder leaving their right arm and shoulder completely uncovered. This tribe was almost completely killed off at the beggining of the twentieth century with the colonization of this area, in order to take over the territory they lived in. There are only a few mestizos left today.

The Yamanas or Yaganes Indians lived in the canals in the southern part of Chile where the tip of South America plunges into the Antarctic Ocean. The Yamanas or Yaganes regularly wore no clothes. The only time they used any type of clothing was during snowstorms, and this was a small square of animal skin, fourty by fourty centimeters, used to cover the part of the body facing the wind. It is possible there are a few of these Indians left in the area of Punta Arenas in Chile (page 22).5

The third tribe that I was referring to is the Alacalufes Indians. It is believed that there are about twenty Alacalufes still living in the southern part of Chile in the Aysen region near Puerto Eden. The Alacalufes never used any clothing at all, not even animal skins or fur. They lived in canoes, and their entire life's activities revolved around the use of a canoe; they spent most of their time in the ocean, fishing and hunting seals and an occasional whale. When the missionaries and Spanish colonists attempted to better their living conditions by giving them clothes and showing them how to make clothes, they inadvertently caused their demise. Apparently, the fact that these clothes would get wet and stay wet from rain, sea water, and snow caused most of the Alacalufes to die from pulmonary congestion (page 22), 6 in addition to diseases introduced by the colonists.

The physical differences developed by these tribes due to their living conditions did not include a coat of hair. The only difference found in the bodies of the Alacalufe (I did not have the opportunity to work with the other two tribes), due to living in the cold, is that there is greater subcutaneous vascularity, which allows more blood to be present in the skin and hence affords greater warmth. These Indians developed a body adapted to their lives in canoes, in that their legs were slightly atrophied while their upper bodies were highly developed due to continuous rowing. If the principal function that hair offered our ancestors was to protect them from the cold, this coat of hair would persist in these Indian tribes, due to their great need.

Darwin says (page 439): *"The view which seems to me most probable is that man, or rather primarily woman, became divested of hair for ornamental purposes, as we shall see under sexual selection; and, according to this belief, it is not surprising that man should differ so greatly in hairiness from all other primates, for characters, gained through sexual selection, often differ to an extraordinary degree in closely related forms."*

Ornamental purposes? To me, this does not make sense. Do we think that a gorilla finds the female less attractive because she is hairy? Ornamental feelings are a subjective judgment. If human females would be hairy, we would find them very attractive, indeed. Porcupine males find their females attractive.

It has been mentioned in some anthropological works that the fact that man does not have hair, demonstrates neoteny, in that the infant does not have hair and an adult retains this characteristic throughout his lifetime. This reasoning seems logical enough, but it does not really answer why this characteristic should be retained. The explanation offered here to this question is, in fact, the simplest and hopefully the most logical a product of biology.

Mammals and birds exhibit either nidifugous or nidicolous characteristics at birth. The nidifugous animals are born with perfect vision, the capability of walking and even running, and with their bodies covered with hair like the adults. The nidicolous animals are born without vision, with limited ability to coordinate movements (except for the most rudimentary), and are born hairless (page 22).7

It was mentioned in the chapter *Man the Only Nidicolous Primate* that man is the only nidicolous primate. If he is, in fact, nidicolous, then he must exhibit those characteristics that would define him as such at birth: lack of vision, little voluntary coordination, and hairlessness, and this is the case. The lack of hair, therefore, is an infantile characteristic. In other nidicolous animals, these characteristics disappear as the animal grows; it develops sight, and the ability to coordinate voluntary movements, and it begins growing hair. In some cases this hair even changes color and quality as the animal reaches adulthood (page 22).8

In human beings, this infantile characteristic is retained, and only small changes in the amount and quality of hair are noted during puberty, as mentioned before. These changes in the amount of hair in the pubic area, armpits, and, it the case of men, on the face and chest with a change of quality and amount of hair on the legs, are part of the development of secondary sexual characteristics (page 22),9 connected with the production of certain hormones.

In human beings, it appears that this aspect of neoteny is not complete, in that upon arriving at puberty, partial changes do occur in only certain areas of the body, leaving others in a neotenic state throughout adulthood. This special neotenic state of man and the proposed explanation here need to be researched further to arrive at a more complete understanding (page22).10

Notes

1. *Man, the Paradoxical Primate*, Jordi Fuentes, 1985
2. *Eschrichtt, Uber die Richtung der Haare am Menschlichen Körper, Müller's Archiv für Anat. und Physiology.* 1837, S. 47.
3. Paget, *Lectures on Surgical Pathology,* 1853, vol. i, p. 71
4. *Man, the Paradoxical Primate*, Jordi Fuentes, 1985
5. Ibid.
6. Ibid.
7. Ibid.
8. Ibid.
9. Ibid.
10. Ibid.

X

The Upright Man

Both *humanoids* and early *Homos* have been frequently represented as hairy beings with greater similarity to gorillas than to human beings, and walking on two feet with their body leaning forwards. As our ancestor became more "humanized," he was represented each time more and more erect and less hairy, until arriving at *Homo sapiens*, at which time he is represented in a totally upright position (page 13).1

This representation of the hominid leaning forward while walking on two feet is just fantasy and physically impossible. This idea was presented as a fast solution to a difficult problem, by someone who did not take time to throroughly reflect upon it. Nevertheless, it has been picked up and repeated by many scientists, including Darwin. If they would have only reflected a little on this subject, they would have seen that this ideas is not possible. When four-legged animals assume a two-legged position, they assume a completely upright position. If they were to do otherwise, their center of gravity would prevent them from maintaining their balance. A hominid walking in an inclined position would be forced to be continually running, because if he stopped for any length of time in this position, he would fall on his face. The only

Representation of the Hominid evolution often used to emphasize humanization, but physically impossible, and in disagreement with evolutionary solutions to the upright position in animals.

way any animal could maintain this stationary position would be if this animal had developed tremendous gluteal muscles. In the cases of the modern-day kangaroo and the bipedal dinosaur, such as the *Tyrannosaurus rex*, the problem of balance was solved by creating a heavy tail as a counterweight. This would have been a possible solution to maintaining balance for man, but the simpler solution was to maintain a completely upright position (page 13), 2 placing the center of gravity in a central vertical axis of the body, with no need for developing a tail. On the contrary:

The tail, instead of being an asset, would constitute a nuisance without function.

The bipedal position is not exclusively human. Birds are also bipedal (page 13),3 and probably birds do not have long tails for the same reason that humans don't have one. In a vertical position of the body, the tail is a nuisance. Yes, yes; I know that birds have a tail. The peacock is one example. But I am referring to a skeletal one. Do you remember the *Archaeoptheryx* the first animal with feathers? Well, the major doubt in considering this creature a bird is the possession of a long bony tail. Another difference was the possession of teeth. I will refer again too this animal at the end of this chapter.

The center of gravity for birds is governed by the position of the feet. The more posterior the feet are with respect to the body, the more erect the bird is forced to walk. The best example of this is the penguin.

Man's legs are found at the bottom of his body (as are the penguins), and for this reason he is forced to adopt an upright position.

With regard to the upright position of the human body, Darwin says (page 433 and 434): *"As soon as some ancient member in the great series of the primates came to be less arboreal, owing to a change in its manner of procuring subsistence, or to some change in the surrounding conditions, its habitual manner of progression would have been modified: And thus it would have been rendered more structurally quadrupedal or bipedal. Baboons frequent hilly rocky districts, and only from necessity climb high trees;* 4 *and they have acquired almost the gait of a dog. Man alone has become a biped; and we can, I think, partly see how he has come to assume his erect attitude, which forms one of his most conspicuous characters. Man could not have attained his present dominant position in the world without the use of his hands, which are so admirably adapted to act in obedience to his will. Sir C. Bell* 5 *insist that "the hand supplies all instruments, and by its correspondence with the intellect gives him universal dominion." But the hands and arms could hardly have become perfect enough to have manufactured weapons, or to have hurled stones and spears with a true aim, as long as they were habitually used for locomotion and for supporting the whole weight of the body, or, as before remarked, so long as they were especially fitted for climbing trees. Such rough treatment would also have blunted the sense of touch, on which their delicate use largely depends. From these causes alone it would have been an advantage for man to become a biped; but for many actions it is indispensable that the arms and the whole upper part of the body should be free; and he must for this end stand firmly on his feet. To gain this great advantage, the feet have rendered flat; and the great toe has been peculiarly modified, though his has entailed the almost complete loss of its power of prehension. It accords with the principle of the division of physiological labour, prevailing throughout the animal kingdom, that as the hands became perfected for prehension, the feet should have become perfected for support and locomotion. If it be an advantage to man to stand firmly on his feet and to have his hands and arms free, of which, from his pre-eminent success in the battle of life, there can be no doubt, then I can see no reason why it should not have been advantageous to the progenitors of man to have become more erect or bipedal. They would thus have been better able to defend themselves with stones or clubs, to attack their prey, or otherwise to obtain food."*

Darwin continues with this exposition concerning to gradual erect posture of our ancestors (page 435): *"As the progenitors of man became more and more erect, with their hands and arms more and more modified for prehension and other purposes, with their feet and legs at the same time transformed for firm support and progression, endless other changes of structure would have become necessary. The pelvis would have to be broadened, the spine peculiarly curved, and the head fixed in an altered position, all of which changes have been attained by man. Professor Schaaffhausen,* 6 *maintains that ""the powerful mastoid processes of the human skull are the result of his erect position; "These processes are absent in the orang, chimpanzee, &c., and are smaller in the gorilla than in man. Various other structures, which appear connected with man's erect position, might here have been added."*

It is not possible to accept the gradual "humanization" that goes hand in hand with the gradual assumption of a vertical or upright position from a quadrupedal or horizontal position. These intermediate positions between the quadrupedal horizontal position and the bipedal vertical position are mechanically impossible to assume while walking. The chimpanzee will occasionally assume this position when he has both hands full with a load of fruit, for example, and he only does this for a few seconds before reverting to the quadrupedal position (page 13).7 When four-legged animals assume an erect position, such as squirrels or prairie dogs and all the members of this large family, as do the simians, meerkats, bears, and Arctic hares, they stand in a completely upright position. The gibbons are one of the anthropoids that are often compared to man; when gibbons walk, they do so on two feet in a totally erect position and use their long arms to maintain balance, with their body directly above their legs (page 14).8

In his attempts to demonstrate the evolution of man for a quadruped to biped Darwin says (page 441): *"Finally, then, as far as we can judge, the tail has disappeared in man and the anthropomorphus apes, owing to the terminal portion having been injured by friction during a long lapse of time; the basal and embedded portion having been reduced and modified, so as to become suitable to the erect or semi-erect position."*

Regarding this point, Darwin did not realize that with a completely upright position, the counterbalance of a tail would be not only useless but detrimental. If we consider the situation, we'll see that it is not necessary to "injure" that tail by "friction" "during a long lapse of time" as Darwin says, to make it disappear. According to evolution, it is only necessary to become useless, and natural selection would take care of its oblivion. No other parts of animal bodies that have been eliminated for lack of use have been "injured by friction" to reach this state of disappearance. This comment is a flaw in Darwin's reasoning. Many animal species have no tails, and did not lose them through injury by friction.

In an erect or bipedal position, the tail would interfere extraordinary with the legs when walking; in addition, it would be a nuisance for the evacuation of the bowels. The tail would become a useless and noxious appendage without a practical purpose, and, probably by Lamarkian action, would disappear. Let us consider that most terrestrial species of primates do not have a tail, or it is rudimentary. The primates that have a long tail are mostly arboreal species that use the tail as a counterbalance or as a rudder to help in long jumps. Those species with a tail that occasionally walk on the ground such as the lemurs and some simian species, have voluntary control over their tail, keeping it erect and well over the ground while walking.

The most probable situation with regard to the tail, is that the most remote ancestors of man had no tail. There are many species of quadrupedal animals that do not have a tail. On the other hand we see that birds are also bipedal, and is not difficult to see that they have no tail either.

a

b

Point of balance

a) Chimpanzee in an upright position. The center of gravity is not vertical, therefore rendering this position unstable and forcing the chimpanzee to use his arms for balance. b) Chimpanzee in a quadrupedal position.

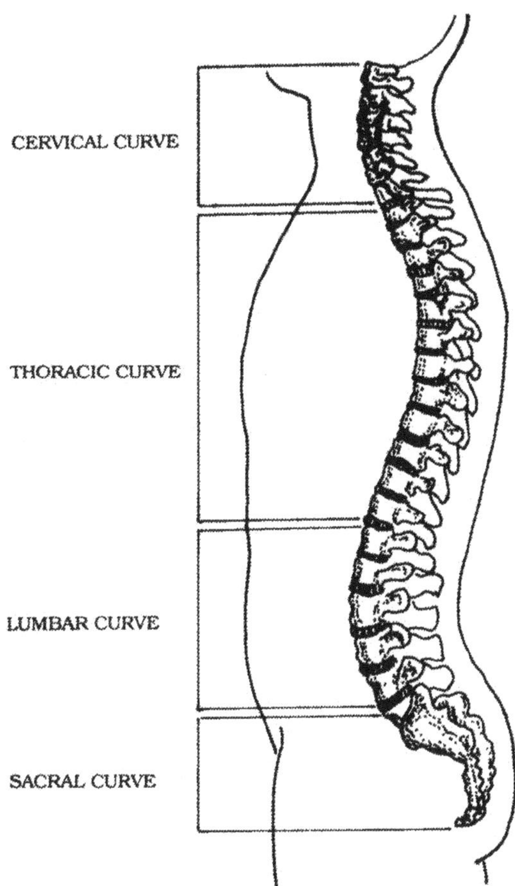

Vertebral column curves

From a lateral view the vertebral column shows four curves. These curves are very important since, like the curves in a long bone, they increase its strength, help to maintain balance in the upright position, absorb shocks from walking, and help protect the column from fracture.

The four curves are alternately convex and concave. The human fetus presents only one concave curve that takes the length of the column. Around the third month after birth the infant begins to hold its head erect, and the cervical curve is formed. At approximately one year the infant begins to stand and walk and the lumbar curve develops. The cervical and lumbar columns are convex anteriorly, while the thoracic and sacral curves maintain the concavity of the fetus, and are referred as Primary curves being the first two Secondary curves, because they are modifications of the fetal curvature.

A spinal column with four curves, a result of bipedalism, is peculiar to humans and no other animal, including the other primates, possess this characteristic. It would appear to be a mechanical impossibility to slowly develop these characteristic four curves from a quadrupedal position to a bipedal position.

The link between modern birds and their possible ancestors, considered the reptiles (with some doubts at the present), is the fossil of the animal called *Archeornis*. This animal already had developed wings with feathers, and could possibly fly, still had a long tail. This tail and the vestigial teeth are what paleontologists consider the link with its ancestor, a sort of reptile. Unfortunately we do not yet have fossils showing the gradual disappearance of the tail until reaching the bird status.

A tail would be as much a nuisance for flying as it would be for walking on two feet. In flight the tail could involuntarily change the direction of flight making it difficult for the bird to steer properly, so it is possible that natural selection would eliminate those individuals with longer tails and conserve those with shorter ones, until after many generations only the ones without tails survived. Birds today steer their flight with the feathers on the tail which they move to one side or the other by means of a short tail vestige, that it is much easier to control, than a long tail with many vertebrae.

As a consequence of all this reasoning we reach the conclusion that:

A semi-erect position is mechanically impossible. The center of gravity would be displaced in front of the vertical axis of the body, and the structure would be forced to fall down, or otherwise develop a counterweight as has happened with the tail of the kangaroo and the some of the ancient dinosaurs.

Man's ancestors have always assumed a completely upright position while walking on two feet.

Notes

1 'Man, the Paradoxical Primate, Jordi Fuentes, 1985.

69 Owen, Anatomy of Vertebrates', vol. iii.p.71

70 'Quarterly Review'. April 1869,p.392

71 in *Hylobates syndactylus*, as the name expresses, two of the toes regularly cohere; and this, as Mr. Blyth informs me, is occasionally the case with the toes of *H. agilis*, *lar,* and *leuciscus*. Colobus is strictly arboreal and extraordinary active (Brehm, 'Thierleben,' B.i.s.50), but whether a better climber that the species or the allied genera, I do not know, it deserves notice that the feet of the sloths, the most arboreal animals in the world, are wonderfully hook-like.

72 Brehm, 'Tierleben', B.i.s.80

73 "The Hand", 7c. 'Bridgewater Treatise', 1833, p.38

76 'On the Primitive Form of the Skull', translated in 'Anthropology Review', Oct. 1868,p.428 ('Anatomy of Vertebrates' vol.ii, 1866,p.551) on the mastoid processes in the higher apes.

XI

The Cerebralization of Man

Man is the most cerebralized (or brainy) of all anthropoids. We love this idea and we are proud of it. Aren't we?

At birth, the human brain of the present human being, *Homo sapiens sapiens*, weighs an average of 350 grams, and an adult brain weighs an average of 1,450 grams. This is an average because the human adult brain may weigh as little as 1,200 grams and as much as 2,000 grams. The greater or lesser weight in these cases does not indicate higher or lower intelligence (page 13).2 The adult gorilla brain weighs approximately 540 grams while the adult chimpanzee brain weighs approximately 380 grams, or slightly heavier than a human newborn brain (page 13).2

If we examine the measurements that have been taken from various hominid skulls that have been found, we find that the cranial cavities increase in size the closer they are to the present time. For instance, the *Australopithecus robustus* at about three million years ago has a cranial capacity in cubic centimeters of about 450 to 500 cc. At about 2.5 million years ago, cranial capacity jumps to 750 cc, and with the appearance of *Homo habilis* two million years ago, cranium capacity has been measured at about 800 cc. A million and a half years ago, *Homo erectus* cranial capacity is measured between 900 and 1,200 cc, nearly modern day man's capacity. A half a million years ago, *Homo sapiens* measured between 1,250 and 1,550 cc capacity. There is little difference between Neanderthal and Cro-Magnon, both having cranial capacities between 1,000 and 2,000 cc. (page 24).2

With reference to the cerebralization of the human being, Darwin inserts in his writings the following paragraph (page 24): *"In the interesting article just referred to ['Les Sélections,' M. P. Broca, "Revue d'Anthropologies, 1873] professor Broca has well remarked that, in civilized nations, the average capacity of the skull must be lowered by the preservation of a considerable number of individuals, weak in mind and body, who would have been promptly eliminated in a savage state. On the other hand, with savages, the average includes only the more capable individuals, who have been able to survive under extremely hard conditions of life. Broca thus explains the otherwise inexplicable fact that the mean capacity of the skull of the ancient Troglodytes of Lozére is greater than that of the modern Frenchmen."*

First, we do not know what Darwin and Prof. Broca understood by Troglodytes, as this adjective means only "cave dweller." Neanderthals and Cro-Magnons were Troglodytes, and their cranial capacity is no different than that of the modern man. When they referred to the Troglodytes of Lozére, they refer to a small quantity of fossil material that cannot be compared to that of a large population. But it is logical to reason that if the more capable individuals among the Troglodytes were the ones who survived, due to their ability to reason, they were the only ones to procreate and perpetuate the larger capacity of the skull. That means that the Troglodytes population as a whole possessed a larger cranial capacity than modern Frenchmen. If this should be the actual case, it does not mean that the Troglodytes were more intelligent than modern Frenchmen, as intelligence is not related to cranial capacity in a given population. Otherwise, if Troglodytes were more intelligent than the modern Frenchmen, it would be a case of retrogression in the French very improbable according to the theory of evolution.

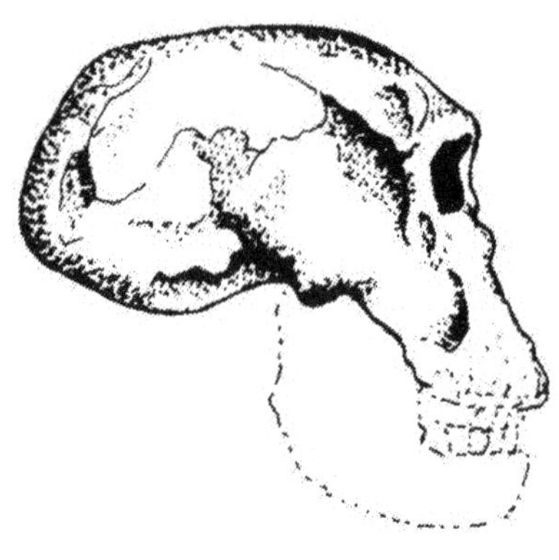

Homo habilis

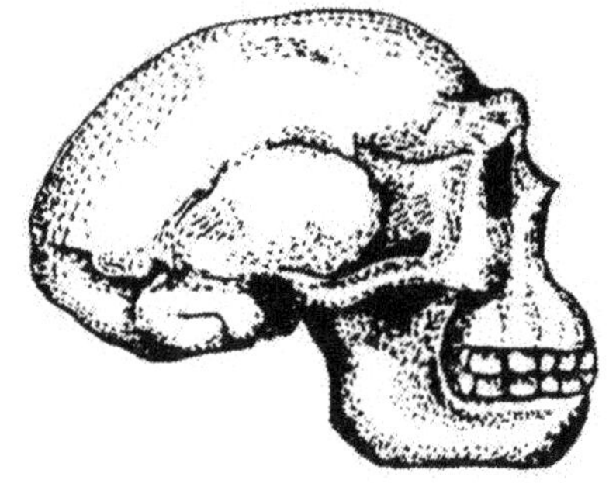

Homo sapiens

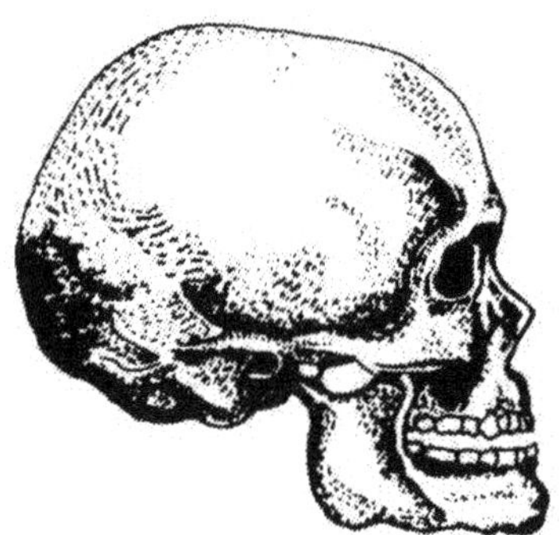

Homo erectus

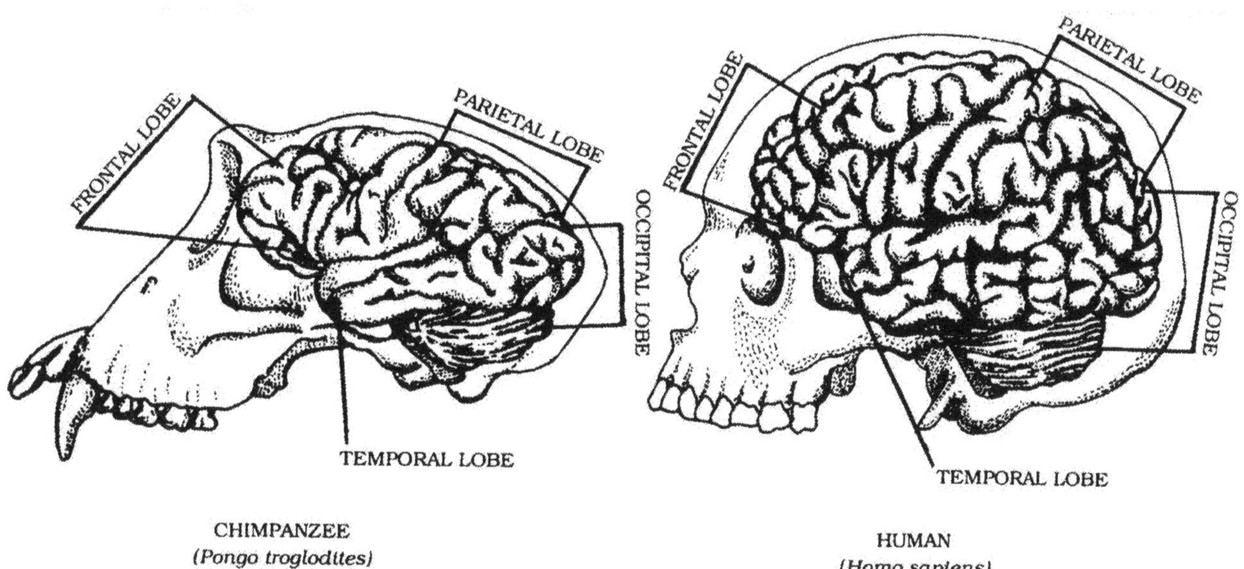

Human brain compared with the chimpanzee brain
The cerebral cortex is more convoluted in man and the frontal and temporal lobes are larger.

If we make an attempt to compare the technical advances with the cranial capacities, we will see that there has been nothing found so far to indicate any technical advances until 750 to 800 cc in capacity was reached. It appears that it was first in *Homo habilis* where we find the intentional making and using of tools. Most of the tools found during *Homo habilis'* time are lithic, since traces of most any other material would have long since perished. Scientists say that at less than 750 cc of cranial capacity, it appears there is not an ability to make stone tools. It is very likely that prior to this, bones, branches, and sticks were used. But here appears to be a contradiction. We can observe in the chimpanzee that, with a cranial capacity of 380 cc, it is able to fashion a tool for extracting termites from termite hills or from trees by denuding a branch of its leaves and using the tool to probe in termite holes (page24).2 Is that the only use a chimpanzee makes of a tool? It is one thing to *use* a tool, and another to *make* a tool. Chimpanzees also use stones to crack open nuts. One stone is used as an anvil, and the other as a hammer. There are isolated cases in which a chimpanzee intentionally hits a stone with another, to produce a chip of stone to be used as a knife. Isn't that tool making?

If a chimpanzee with a cranial capacity of 380 cc can intentionally make a stone chip to use as a knife, there is no reason to ignore the possibility that with 750 cc cranial capacity, the humanoids before *Homo habilis* could also make stone tools, as chimpanzees do with less capacity.

Darwin makes the following comment (page396): *"Vulpian 5 remarks," Les différences réelles qui existent entre l'encéphale de l'homme et celui des singes supérieurs, sont bien minimes. Il ne faut pas se faire d'illusions à cet égard. L'homme est bien plus près des singes anthropomorphes par les caractères anatomiques de son cerveau que ceux-ci ne le sont non seulement des autres mammifères, mais méme de certains quadrumanes, des guenons et des macaques." But it would be superfluous here to give further details on the correspondence between man and the higher mammals in the structure of the brain and all other parts of the body."*

Scientists sometimes seemingly like to complicate things, and the most logical explanations are often ignored or overlooked. Why? Why do we have to search for difficult, twisted, and mysterious explanations instead of dealing with, studying, and accepting simple and logical explanations? Do the scientists feel degraded in their expertise if they accept an openly logical and simple solution? Or perhaps their professional zeal impels them to reach the most unreachable conclusions. Anyway, professional jealousy and rivalry between scientists can play an active role in the exposure of ideas, which may divert them from the central point of thought.

The making of more perfect stone tools requires more advanced tool-making methods, not just hitting one stone with another. However, it is very unlikely that stone tools were the first tools used by the predecessors of hominids, except for stones used in their natural state without further elaboration. In order to make a stone tool not only would the individual have to know the kind of stone to use, but also the angle of impact upon fashioning the blade, in order to produce a usable cutting edge. The ability to make these stone tools repeatedly and intentionally does not seem to appear until reaching a cranial capacity double that of the chimpanzee's.

So far there have not been enough australopithecine fossils found between four and three million years old to enable us to determine their state of technical advancement, taking into account their cranial capacity, and hence allowing us to deduct the possibilities of their brain functions.

The cerebralization of man appears to have originated parallel with the slow development of the human embryo body and dramatic brain development during gestation. At birth, a newborn's brain is highly developed, although its capacity to enable independent functioning is still incomplete. At birth, the infant is not prepared to process all the messages that his environment delivers to him and to react appropriately. Only a few involuntary reflexes are present. But even though the brain has all the necessary elements to eventually react appropriately, voluntary reactions whereby he consciously processes his environment, understands it, and then has his body react accordingly, takes time to develop approximately a month for vision, a year for balance, and for a boy, two years for speech. Girls speak and walk earlier, on average, than do boys. There are exceptions, as biology is not an exact science. (pag.25).2

The brain is also where the ability to originate abstract ideas and transmit them to others through another cerebral activity called language lies. Language is produced by creating acoustic waves in vocal cords that are modulated in the mouth and lips, producing the thousands of different sounds of which *Homo sapiens* is capable (pag.25).2

With regard to this subject of the cerebralization of man, Darwin, in the chapter, *"Natural Selection,"* makes this large exposition (page 430 and 431): *"We have now seen that man is variable in body and mind; and that the variations are induced, either directly or indirectly, by the same general causes, and obey the same general laws, as with the lower animals. Man has spread widely over the face of the earth, and must have been exposed during his incessant migration to the most diversified conditions. The inhabitants of Tierra del Fuego, the Cape of Good Hope,and Tasmania in the one hemisphere, and the Arctic regions in the other, must have passed through many climates, and changed their habits many times, before they reached their present homes.* 64 *The early progenitors of man must also have tended, like all other animals, to have increased beyond their means of subsistence; they must, therefore, occasionally have been exposed to a struggle for existence, and consequently to the rigid law of natural selection. Beneficial variations of all kinds will thus, either occasionally or habitually, have been preserved, and injurious ones eliminated. I do not refer to strongly-marked deviations of structure, which occur only at long intervals of time, but to mere individual differences."* Here, Darwin already realized the existence of subtle changes, described by Niles Eldredge as *"Punctuated Equilibria"*.9

Darwin continues (page 431 and 432): *"We know, for instance, that the muscles of our hands and feet, which determine our powers of movement, are liable like those of the lower animals,* 65 *to incessant variability. If then the progenitors of man inhabiting any district, especially one undergoing some change in its conditions, were divided into two equal bodies, the one half which included all the individuals best adapted by their powers of movement for gaining subsistence, or for defending themselves, would on an average survive in greater numbers, and procreate more offspring than the other and less well endowed half. Man in the rudest state in which he now exists is the most dominant animal that has ever appeared on this Earth. He has spread more widely than any other highly organized form, and all others have yielded before him. He manifestly owes his immense superiority to his intellectual faculties, to his social habits, which lead him to aid and defend his fellows, and to his corporeal structure. The supreme importance of these characters has been proved by the final arbitrament of the battle for life. Through his powers of intellect, articulate language has been evolved; and on this his wonderful advancement has mainly depended. As Mr. Chauncey Wright remarks,* 66 *'a psychological analysis of the faculty of language shews that even the smallest proficiency in it might require more brain power than the greatest proficiency in any other direction.' "He has invented and is able to use various weapons, tools, traps, etc. with which he defends himself, kills or catches prey, and otherwise obtains food. He has made rafts or canoes for fishing or crossing over the neighbouring fertile islands. He has discovered the art of making fire, by which hard and stringy*

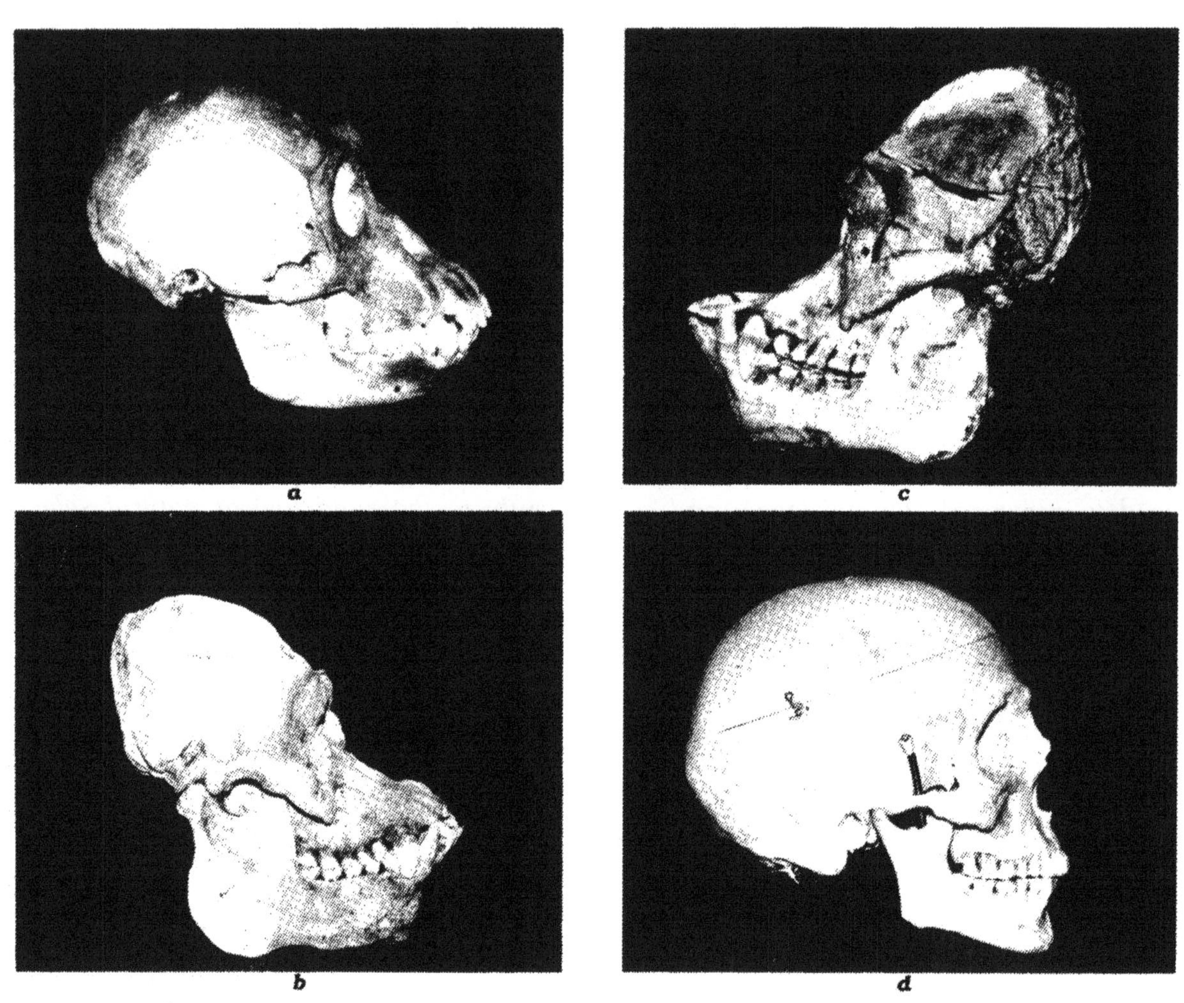

Lateral view

a) chimpanzee skull b) gorilla skull

c) orangutan skull d) human skull

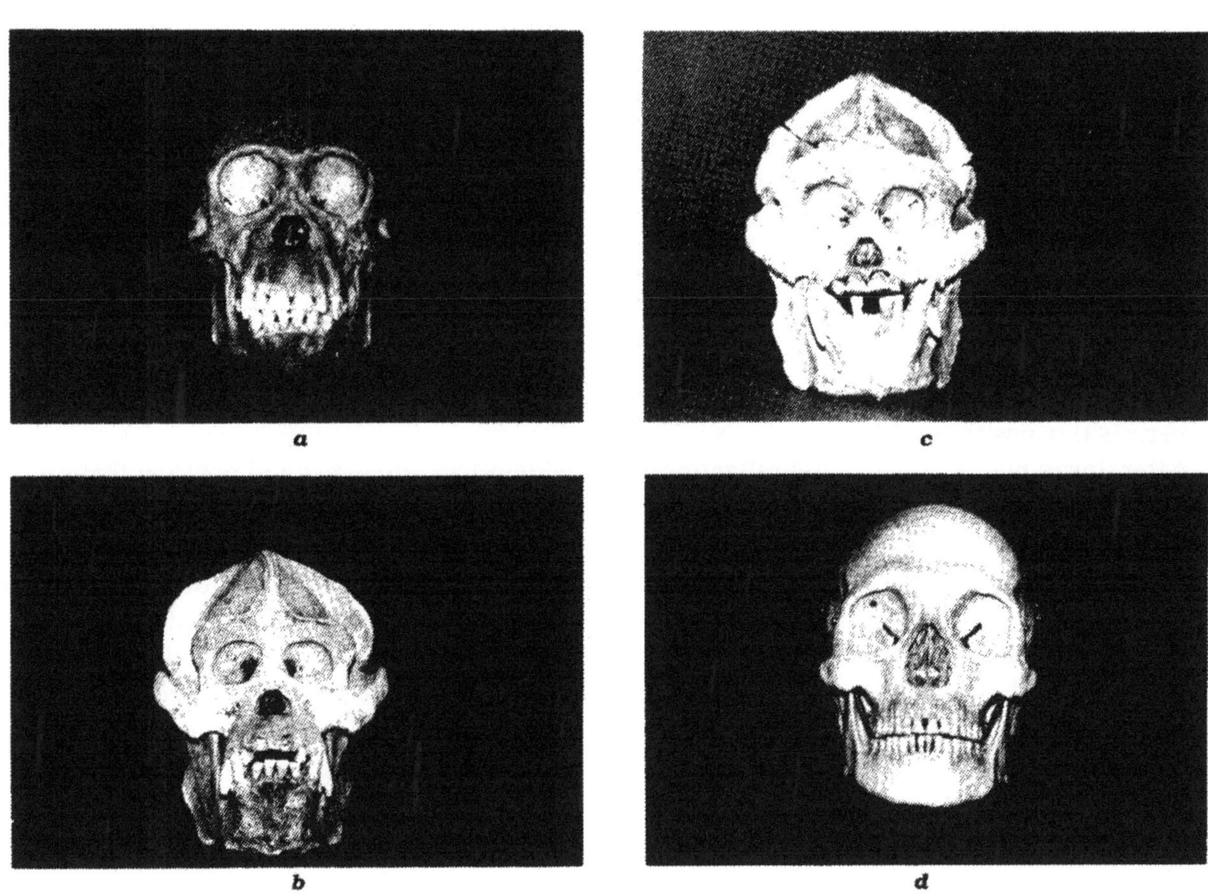

Frontal view of skull
a) chimpanzee b) gorilla c) orangutan d) human

roots can be rendered digestible, and poisonous roots or herbs innocuous. This discovery of fire, probably the greatest ever made by man, excepting language, dates from before the dawn of history. These several inventions, by which man in the rudest state has become so pre-eminent, are the direct results of the development of his powers of observations, memory, curiosity, imagination, and reason. I cannot, therefore, understand how it is that Mr. Wallace 67 *maintains that 'natural selection could only have endowed the savage with the brain a little superior to that of an ape.'"*

At this point, Darwin considers language a product of discovery. I disagree: Language evolved in a human being, but the human being did not discover it. In order to discover something, this something has to exist previously. Language evolved gradually, fast or slow, but from zero to a state where it could be used to communicate ideas. This is not discovery; this is development. Language can be developed, but not discovered, as we can discover blood circulation or the digestive processes.

Cerebralization is considered to be the reason for our advanced state of technology in all areas. It is the cause of the development of philosophy, art, religion, and all the expressions of creativity and expansion of abstract thoughts and ideas. In other words, cerebralization is seen as the acme of human ascension to the highest level of evolution involving esoteric disciplines.

Seen from this point of view, cerebralization could be considered a positive feature. But . . . is it? Where is the human species going with the results of our cerebralization, if, with it, we have caused the destruction of the ozone layer, the pollution of the air, soil, water, forests, and oceans, and the extinction of many species of animals and plants? All we have destroyed is gone forever. Is this positive?

After this exposure, could cerebralization be considered an advanced and positive feature? In order to be positive, it should be used for the conservation of the spaceship on which we travel through the universe, the planet Earth. This is the place where we live and where we are supposed to continue living; but destruction such as we are causing does not bode well for the future. Where are we going with our cerebralization, which caused us to separate from nature, ignore its laws, and created in us the idea of being demigods? The future of the human species that has evolved from lower forms of life to the present state does not look very promising, since the same cerebralization has created in man the appearance of negative features such as selfishness, greed, lack of respect for other creatures, the use of aggression to obtain our goals, mass destruction of our own species, and so on.

Is cerebralization of man nature's aberration? Let us think about it. The most optimistic individuals say that the same cerebralization will enable us to improve the present situation and repair the damage caused by our civilization. Man has the faculty to correct himself and correct past mistakes. Very poetic. But is there any time left to correct so many mistakes? For some species, there is no return; they are extinct forever. What about ours?

The world does not grow, but the human species does. Will there be enough room for all the humans, including enough space to produce food for all of them? Let us reconsider the ideas of Thomas Robert Malthus. 68 His ideas suggested to Charles Darwin the relationship between progress and the survival of the fittest. Some thinkers can say that Malthus' prediction that population will increase more rapidly than food supplies failed because cerebralization allowed man to improve the production of food by using better methods in agriculture; but as many conservationists warned, production could not keep pace with population indefinitely.

Where is the glory of man's cerebralization? This is food for thought.

Notes

1. *Man the Paradoxical Primate*. Jordi Fuentes, 1985.
2. Ibid.
3. Ibid.
4. Ibid.
5. *"Leç. Sur la Phys."* 1866, p. 890, as quoted by M. Dally. *l'Ordre des Primates et le Transformisme*, 1868. p. 29
6. Man, *the Paradoxical Primate*. Jordi Fuentes, 1985.
7. Ibid.
8. Latham, *Man and his Migrations*, 1851, p. 135.
9. *Time Frames*, Niles Eldrege, Simon and Shuster, 1985.
10. Messrs. Murie and Mivart, in their "Anatomy of the Lemuroidea *"Transactions of the Zoological Society"*, vol. vii, 1868, p. 96-98), say "Some muscles are so irregular in their distribution that they cannot be well classed in any of the above groups." These muscles differ even on the opposite sides of the same individual.
11. "Limits of Natural Selection," *North American Review*, Oct. 1870, p. 295.
12. *Quarterly Review*, April 1869, p. 392
13. "Essay on the Principle of Population," 1798.

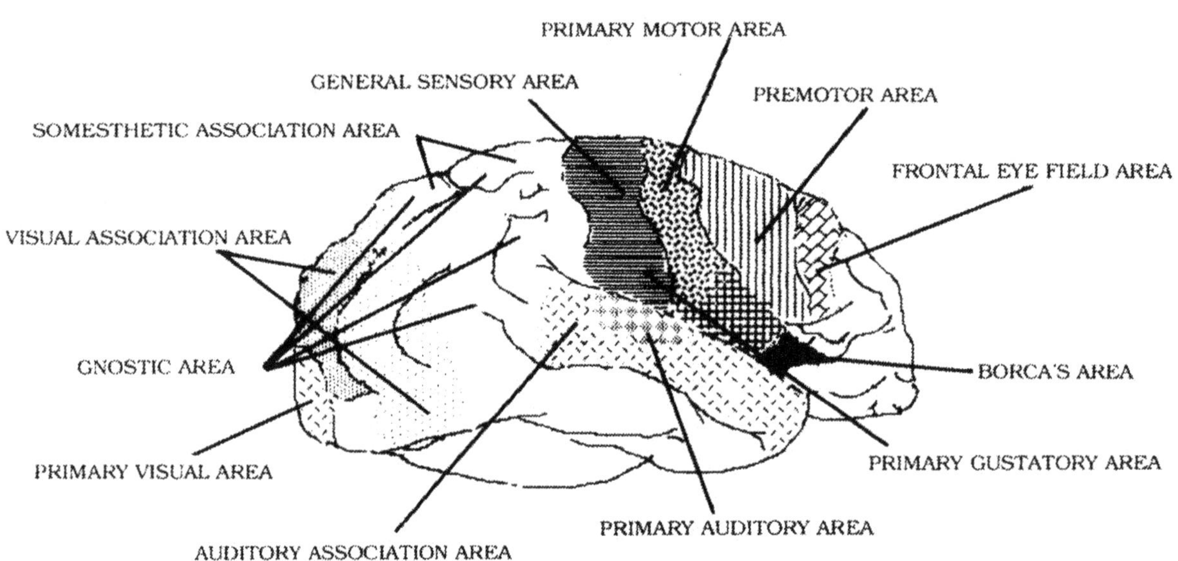

Functional areas of the cerebrum. Although the right hemisphere is illustrated, Broca's area is in the left hemisphere of most people.

Functional Areas of the Cerebrum

1. General sensory area. Located directly posterior to the cerebral sulcus of the cerebrum in the postcentral gyrus of the parietal lobe. It receives sensations from cutaneous, muscular, and visceral receptors in various parts of the body. The major function is to localize exactly the points of the body where sensations originate.

2. Somesthetic association area. Located immediately posterior to the general sensory area. This area permits you to determine the exact shape and texture of an object without looking at it. It also stores memories of past sensory experiences.

3. Primary visual area. Located on the medial surface of the occipital lobe and occasionally extends around to the lateral surface. It receives sensory impulses from the eyes and interprets shape and color.

4. Visual association area. Located in the occipital lobe. It receives sensory signals from the primary visual area and the thalamus. It relates present to past visual experiences with recognition and evaluation of what is seen.

5. Primary auditory area. Located in the superior part of the temporal lobe near the lateral cerebral sulcus. It interprets the basic characteristics of sound, such as pitch and rhythm.

6. Auditory association are. Inferior to the primary auditory area in the temporal cortex. It determines if a sound is speech, noise, or music. It interprets the meaning of the speech by translating words into thoughts.

7. Primary gustatory area. Located at the base of the postcentral gyrus, above the lateral cerebral sulcus in the parietal cortex. It interprets sensations related to taste.

8. Primary olfactory area. Located in the temporal lobe on the medial aspect. It interprets sensations related to smell.

9. Gnostic area. Located between the somesthetic visual and auditory association areas. It receives impulses from these areas and from the taste and smell areas, the thalamus, and lower portions of the brain stem. It integrates sensory interpretations from the association areas and impulses from other areas so that a common thought can be formed the various sensory inputs. At this point, it transmits signals to other parts of the brain to cause the appropriate response to the sensory signals.

10. Primary motor area. Located in the precentral gyrus of the frontal lobe. It controls specific muscles or groups of muscles. Stimulation of a specific point of the primary motor area produces a muscular contraction, usually on the opposite side of the body.

11. Premotor area. Anterior to the primary motor area. It is concerned with learned motor activities. It generates impulses that cause a specific group of muscles to contract in a specific sequence; thus, it controls skilled movements.

12. Frontal eye field area. Located in the frontal cortex. It controls voluntary scanning movements of the eyes.

14. Broca's area. Located in the frontal lobe just superior to the lateral cerebral sulcus. It translates thoughts into speech, sending signals to the premotor regions that control the muscles of the larynx, pharynx, and mouth. It also sends signals to the primary motor area, and from there impulses are sent to regulate the proper flow of air past the vocal cords. The coordination of these impulses and the muscular contractions they produce is what converts the thoughts into speech.

XII

The Beggining of Language

One of the most notable differences between man and ape, and between all other animals, is man's ability to communicate by the use of spoken language (page 26).1 At least, that is what was thought until now. Today we know that many animals have a language and communicate with members of the same species. Even the existence of family dialects has been demonstrated. These dialects are used to communicate between members of the same family. One well-known example is the killer whales (*Orcinus orca*), with several family dialects.

Darwin, in his chapter titled *"Language,"* says (page 461): *"This faculty has just been considered as one of the chief distinctions between man and the lower animals. But man, as a highly competent judge, Archbishop Whately remarks, 'is not the only animal that can make use of language to express what is passing in his mind, and can understand, more or less, what is so expressed by another.'* 2 *In Paraguay the Cebus azaroe, when excited, utters at least six distinct sounds, which excite in other monkeys similar emotions."* 3

This remark of Archbishop Whately was prophetic, as it has been proved right. Today we know that some sounds emitted by some animals about danger are understood by animals not only of the same species, but by animals of different species, even of different orders.

Much has been written about the origin or the beginning of language, but up until now no hypothesis is completely satisfactory. It has been theorized that perhaps language began approximately 40,000 years ago and coincided with the cultural explosion of the time. This possibility is something to be considered, but if one reflects long enough about this question there are also sound reasons to refute this position. Biologically, the development of language, with all of its cerebral implications, the mechanics to create vocal cords and special musculature for the flexibility needed by the tongue, lips, and mouth cavity, in only a few thousand years seems highly improbable (page 26).4 fourty-thousand years is only a second in the evolution of man.

Continuing with this subject, Darwin says (page 463): *"Since monkeys certainly understand much that is said to them by man, and when wild, utter signal-cries of danger to their fellows; and since howls give distinct warnings for danger on the ground, or in the sky from hawks (both, as well as a third cry, intelligible to dogs),5 may not some unusually wise ape-like animal have imitated the growl of a beast of prey, and thus told his fellow-monkeys the nature of the expected danger? This would have been a first step in the formation of language."*

At this point Darwin assumes that the first step for the formation of language consisted of the "fellow monkeys," who understood what a "wise ape-like animal" said or pronounced. According to the definition given in *Webster's Dictionary*, an Ape is : "a chimpanzee, gorilla, orangutan, or gibbon; a large tailless monkey that can stand and walk in an almost erect position." "Monkey" is defined in the same dictionary as "any of the primates (the highest order of animals, except man, and usually, the lemurs); specifically any of the smaller, long-tailed members of the primates excluding anthropoid apes."

According to definition, there is a difference between apes and monkeys. Did Darwin understand the names to be synonymous? If he did, we can understand what he meant in the previous paragraph. If he was aware (as we assume) of the difference between both types of animals, as he confirmed many times in his works, this last statement is confusing. Perhaps he meant that one species can understand some sound produced from some other species. If this should be the case, as today it happens with the sounds of alarm produced by some species that are understood by others, do we have to assume, then, that the creation of language was the result of interaction between different species? That is hard to swallow. I am not going to dwell more on a point that seems more a hasty comment than a true scientific thought.

Darwin presents the possibility that *"the first step in the formation of language"* could be the imitation of the growls of a beast of prey to tell his fellow-monkeys the nature of the expected danger. I want to say here that many species of monkeys have different sounds for different dangers, sounds that all of the other members of the species understand, and none of them is the "imitation of the growl of a beast of prey." Most of the languages have onomatopoeic sounds, but considering this to be the origin of language seems to go too far.

Darwin continues developing his ideas about the formation of language, and states (page 463): *"As the voice was used more and more, the vocal organs would have been strengthened and perfected through the principle of the inherited effects of use; and this would have reacted on the power of speech. But the relation between the continued use of language and the development of the brain has no doubt been far more important. The mental powers in some early progenitor of man must have been more highly developed than in any existing ape, before even the most imperfect form of speech could have come into use."*

I agree 100 per cent with Darwin on this point. Speech is a product of cerebralization, and I understand, by cerebralization, the augment in volume and hence in mental powers of the brain. Before developing language, the brain had to be capacitated for developing such activity; but it is very possible that mental capacity and initiation of language developed simultaneously during a very long period of time.

Following the development of this subject, Darwin states (page 462): *"Moreover, no philologist now supposes that the language has been deliberately invented; it has been slowly and unconsciously developed by many steps."* 6

Darwin was absolutely right concerning this point. Language was not invented as a conscious idea or project, but was developed as a consequence of the need to communicate between individuals. This development was achieved through a very long span of times, as the evolution of the organs of speech and the area in the brain that could process this activity could not be produced in a short time; on the contrary, the process to develop both areas simultaneously has to be achieved during an enormous span of time, probably millions of years.

Such a complex mechanism cannot simply appear in a relatively short time; we have to consider that the zone in the brain associated with speech was already well developed more than a million years ago (page 26),7 as has been established with the study of the skulls of this period.

Following the development of this point, Darwin says (page 465): *"As all the higher mammals possess vocal organs, constructed on the same general plan as ours, and used as a means of communication, it was obviously probable that these organs would be still further developed if the power of communication had to be improved; and this has been affected by the aid of adjoining and well adapted parts, namely the tongue and lips."*

I agree that most of the "higher mammals" possess vocal cords. Where I object is that not only the tongue and lips contribute to articulate language. Darwin does not take into account two of the most important features in the articulation of sounds once produced by the vocal cords. These are the muscles of the face and the shape of the mandible and uniformity of tooth size.

Here, Darwin ignores the importance of the teeth in the formation of the many different variations of sounds used in the human language. This idea is further developed in the following chapters Why Human Teeth Are Human, and The Human Face.

On the course of developing the chapter about language, Darwin makes the following observation (page 461): *"It is a more remarkable fact that the dog, since being domesticated, has learnt to bark in at least four or five distinct tones.* 8 *Although barking is a new art, no doubt the wild parent-species of the dog expressed their feelings by cries of various kinds. With the domesticated dog we have the bark of eagerness, as in the chase; that of anger, as well as growling; the yelp or howl of despair, as when shut up; the baying at night; the bark of joy, as when starting on a walk with his master; and the very distinct one of demand or supplication, as when whining for a door or window to be opened."*

Wild relatives of the domesticated dog also possess several different cries and sounds to communicate different feelings, even if they do not possess the same bark as dogs do. Coyotes, whose scientific name is *Cannis latrans*, possess, as the species' name implies, a bark that is very characteristic, typical, and, I should say, even harmonic. I know that many people could object to the adjective "harmonic" in the case of the coyote howling. The application of adjectives is, most of the time, a subjective matter. I love coyotes. They are intelligent and resourceful creatures.

Unfortunately, archaeology cannot directly lend much help with respect to language, as there can be no fossil providing the existence of the modality of communication. However, indirect archaeology can shed some light on the subject. Intentional stone tools have been dated back to two million years ago, and these are instruments that were made with premeditation and with a specific goal in mind. Some scientists believe that the first real distinction between man and ape came when man began to fashion stone tools.

The simians, whether they are apes or Old and New World monkeys, also have a limited ability to use and make tools. The capuchin monkeys, belonging to the diverse *Platyrrhini* family of monkeys found in America, also use tools. They use stones as hammers to break open nuts, and they use thick branches as levers (remember that a lever is considered, in physics, a machine, as it produces work) to move weights or to get into holes in trees where they may find insect larvae. Chimpanzees use several tools. They use a stone as a hammer, they use a pinch of leaves as a sponge to soak up water from holes in trees, and they use sticks to obtain termites from their hills or home trees. They also use sticks to fend off leopard attacks. The chimpanzee remains alone in his ability to actually transform a raw material into something more finished to achieve his end. He does this by denuding a branch of its leaves to have a stick he may poke into a hole to obtain termites. There is no doubt that this act is premeditated, with a specific goal in mind. There is a bird that uses the same technique toward the same purpose. Man is the only other animal that is capable of transforming materials into tools with specific ends in mind (page 26).9 Sea otters use tools, but they do not make them; they use stones without any modification.

The first *Homo* tools date back approximately two million years ago, and they have been found in Africa and in Asia. Reflect just for a moment on the implications of this fact. If these tools are found in various parts of the world, it means that there was more than one person able to make them. It is not a phenomenon that would be attributable to an isolated inventor. This phenomenon could have developed simultaneously in different areas, but what couldn't have happened is that each time a tool was made, it was an individual discovery by a man. In each locale, there undoubtedly were several individuals who could make these tools, although the actual inventor was only one. What is the inference from the fact that there were several individuals using the same technique and the same material to make tools? There had to have been some type of communication between these individuals. Imitation would not be an adequate explanation here, because perhaps one individual would try to imitate another, but would soon find himself frustrated when, upon hitting one stone against another, his goal was not realized. To be able to succeed with making tools, he who is learning needs instruction from he who knows. The teacher, in this case, would have to show his student how to choose the proper material. The materials used by these individuals are not usually indigenous to the area, but have been transported. This would most likely indicated that a maker of the tools knows his material and where it must be obtained. Once the student has the proper material, he would need instruction on the angle at which the stone must be hit and how to hit it to achieve the cutting edge desired. By imitation alone his goal would not be reached. There would have to be a language, however rudimentary, to express ideas such as "white, good stone" (for quartz), "black, hard stone" (for basalt), and "gray, bad stone" (for granite). If all of this requires language, it may be deduced that language preceded technique (page 26).10 Language was not created as a result of obtaining tools. Before that, language should have existed to communicate many other ideas relative to the survival of the group. Baboons have different sounds to indicate different dangers. One sound means "snake," and all the members get into a standing position to see where the snake is coming from. Another means "eagle," and as soon as a baboon emits this sound, all members look at the sky to detect the incoming danger. "Leopard" also has its specific sound in baboon language. More than thirty different words in this simian language have been detected. Most species of animals have a phonetic means of communication with specific meanings in their vocabulary.

It is well known, and documented, that whales have language, and that making use of it allows them to communicate with and keep in contact with members of the same species. Elephants and rhinoceroses possess a low-frequency sound language, not perceptible by human ears, that is used to communicate among individuals of the same species at relatively long distances, and in situations when visual communication is not possible. Hippopotamuses have a language with two modalities that permits them to communicate among individuals in and out of the water. One is produced by the nostrils when the animal has its head at the surface of the water. The other is produced under water as a series of clicks produced when the animal moves the enclosed air in its throat, and the communication is received through the jaw bones instead of the ears, which have no use under water.

Language, as we know it for our species, is a product of the intense cerebralization of man. As the cerebral cortex capacity augments, the ability to think abstract ideas also increases. These ideas can only be transmitted through language. Darwin comments (page 461): *"The habitual use of language is, however, peculiar of man; but he uses, in common with the lower animals, inarticulate cries to express his meaning, aided by gestures and the movement of the muscles of the face."* 11 Here, at last, Darwin considers important the muscles of the face in body-language communication, but he does not consider their importance in the production of different sounds.

It would seem that to achieve a high degree of cerebralization, man would need the parallel development of language and, consequently, abstract ideas. It seems logical to me to consider the possibility to create abstract ideas as a function developed long after the creation of language. The reason why is that even today, after thousands and, possibly, millions of years since the beginning of language, still there are cultures in which abstract concepts are unknown. For instance, the natives of New Guinea did not have the concept of zero, and they could only count to five. The introduction of these concepts to their culture and to some more advanced ones happened with the invasion of Western culture, initiated by Australians in the first quarter of this century.

The most elemental conceptual distinctions, such as good and bad, hard and soft, and light and dark, imply the existence of an abstract idea which suggest an enormous advance in the evolutionary scale and in time, and a marked tendency toward specialization not achieved by any other living creature on Earth.

It is relatively evident that technology could not exist without language, but it is not so evident that language could not exist without technology. We can find an example of this in the Tasaday tribe in the Philippines; in spite of having a well developed language in terms of syntax and morphology (but poor in vocabulary), their technology could be compared to hominid technology. This tribe, belonging to *Homo sapiens* has not developed techniques for making clothing, tools, or utensils, but the tribe members are fundamentally social and cooperative. Therefore, in this case, we can plainly see that language can exist for a long time without necessarily giving rise to technology. What is harder to conceive is that technology could exist prior to language (pag.27).12 To develop technology, the exange of ideas is necessary; hence, the use of language.

From this line of thinking, it can be inferred that language has been in existence at least as long as stone toolmaking, in other words, since the time of *Homo habilis*, or two million years ago. It is possible that the existence of language is even older than, that, as we will see in the next chapter. Tool-making is a very specialized technique, so language should precede this activity.

Darwin makes a rather romantic statement about the origin of language when he says (page 463): *"When we treat the sexual selection, we shall see that primeval man, or rather some early progenitor of man, probably first used his voice in producing true musical cadences, that is in singing, as do some of the gibbon-apes at the present day; and we may conclude from a widely spread analogy that this power would have been especially exerted during courtship of the sexes, and would have served as a challenge to rivals."*

I am not going to challenge this possibility, as many animals have special vocalizations used only in the pre-mating or courtship period, and it is possible that this could have been used by our ancient ancestors. We know how vocal deer, elk, and moose are during their runs. What does not seem logical is to attribute to this particular activity the creation of language. Present humans are continuously sexually active animals without a definite period of sexual activity. It is not difficult to admit that, being the present human being, a creature with such particularity, our ancestors would also possess this characteristic; in this case, the "musical cadences" of Darwin should be emitted continuously. Can you visualize a community of several dozens of individuals singing every day and simultaneously expressing their feelings during this chronically continuous courtship? What a concert!

Here, Darwin unconsciously compares man's behavior with the poetic behavior of birds. We know that the song of a bird is, in many cases, advertising the possession of a territory, and not really a way to convince the female to accept the advances of the male. There are many birds that sing even without breeding territories, and they sing also out of breeding season. Every autumn, the songs of the white-crowned sparrow (*Zonoytichia leucophrys*) announce their arrival at Southern California, very far from their breeding grounds, which reach as far as the Arctic tundra. Many animals produce certain sounds during courtship, but too assume that the origin of language had this origin is, to my understanding, to go too far.

The Appearance of Anatomical Details Favoring the Development of Language

1. Vertical position of the body and rotation of the head with respect to the spinal column.
2. Rounded head allowing greater cranial capacity and placement of face in a near-vertical position.
3. A vertical face without prognatism allowing freedom of movement of the facial muscles essential to speech.

4. Enlargement of the cerebral capacity and enlargement of the various areas of the human brain, especially the areas associated with abstract thought and the development of speech.
5. The lack of enlarged canines and the lack of diastema.
6. The arched form of dentition corresponding to the shape of the tongue.
7. The appearance of vocal cords. (Common with other mammals).

Notes

1. *Man, the Paradoxical Primate*, Jordi Fuentes, 1985.
2. Quoted in *Anthropological Review*, 1864, p. 158
3. Rengger, *Anthropological Review*, 1864, s. 45.
4. *Man, the Paradoxical Primate*, Jordi Fuentes, 1985.
5. Houzeau gives a very curious account of his observations on the subject in his *Facultiés Mentales des Animaux*, tom. ii, p. 348
6. See some good remarks on this by professor Whitney, in his *Oriental and linguistic Studies*, 1873, p. 354. He observes that the desire for communication betwen man is the living force, which, in the development of language, "works both conciously and unconciously; consiously as regards the immediate end to be attained; unconciously as regards the immediate end to be attained; uncosciously as regards the further consequences of the act."
7. *Man, the Paradoxical Primate*, Jordi Fuentes, 1985.
8. See my Variation of Animals and Plants under Domestication, vol. i, p. 27.
9. *Man, the Paradoxical Primate*, Jordi Fuentes, 1985.
10. Ibid.
11. See a discussion on this subject in Mr. E. B. Tylor's very interesting work, *Researches into the Early History of Mankind*, 1865, chpaters ii to iv.
12. *Man, the Paradoxical Primate*, Jordi Fuentes, 1985.

XIII

Why Human Teeth are Human

Thinking about the title of this chapter, the reader might say: "Why do you say that? That is a stupid question! They are human because they belong to a human being! Simple, isn't it?" Well, it is not that simple.

When archaeologists and anthropologists discover an ancient tooth, they try to identify the creature to which it belonged. If, after a very meticulous and detailed study and analysis of this material, the results indicate that the tooth resembles present-day human teeth, they consider the possibility that this creature was a human or was in the human line of development. In other words, if the tooth in question looks human, it is classified as human. It seems too easy, but actually it is the result of a profound study of the different characteristics of the subject tooth compared with the teeth of other mammals and man, and their evolution.

Animal teeth have evolved and developed through the centuries in response to the particular eating and chewing needs of the individual species. Or have the eating and chewing needs evolved in response to the development of teeth? Here is the chicken and egg dilemma. Probably both have evolved simultaneously.

In relation to dentition, Darwin makes the next statement (page 407): *"It appears as if the posterior molar or wisdom-teeth were tending to become rudimentary in the more civilized races of man. These teeth are rather smaller than the other molars, as is likewise the case with the corresponding teeth in the chimpanzee and orang; and they have only two separate fangs. They do not cut through the gums till about the seventeenth year, and I have been assured that they are much more liable to decay, and are earlier lost than the other teeth; but this is denied by some eminent dentists. They are also much more liable to vary, both in structure and in the period of their development, than the other teeth.* 1 *In the Melanian races, on the other hand, the wisdom-teeth are usually furnished with three separate fangs, and are generally sound; they also differ from the other molars in size, less than in the Caucasian races.* 2 *Professor Schaffhausen accounts for this difference between the races by noting that the posterior dental 'portion of the jaw being always shortened' in those that are civilized,* 3 *and this shortening may, I presume, be attributed to civilized men habitually feeding on soft, cooked food, and thus using their jaws less. I am informed by Mr. Brace that it is becoming quite a common practice in the United States to remove some of the molar teeth of children, as the jaw does not grow large enough for the perfect development of the normal number."* 4

Some teeth are highly developed in different animals, and for specific purposes. Man once again offers a paradox; he is the only carnivorous primate that has not developed carnivorous teeth or sharp, pointed front teeth, large canines, and serrated molars. All of the other primates, including the anthropoids, possess carnivorous teeth. Gorillas are strictly vegetarians, but still have very characteristic carnivorous teeth with large canines. Why? Are they defensive – offensive weapons?

Human teeth are peculiar to this species. Any animal fossil tooth with "human" characteristics makes a good candidate for classification as a hominid or "human-like."

Human teeth are directly involved in the process of speech, acting as a brace

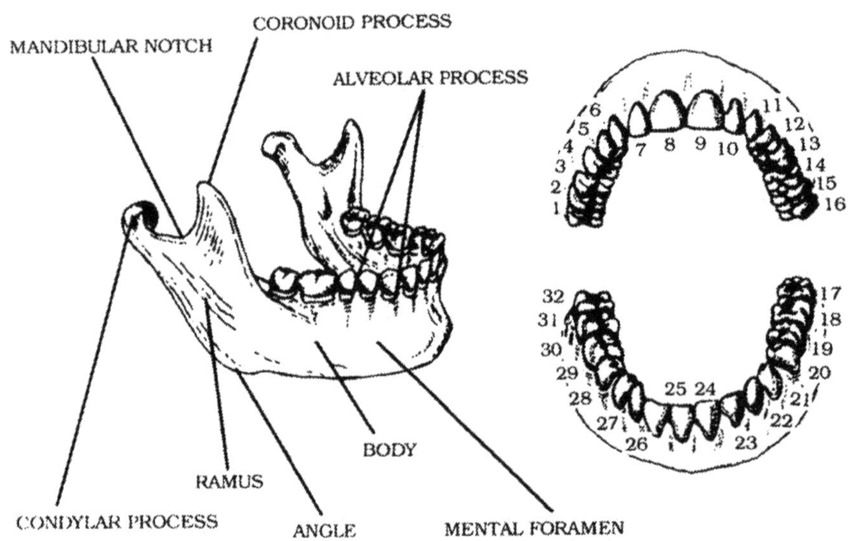

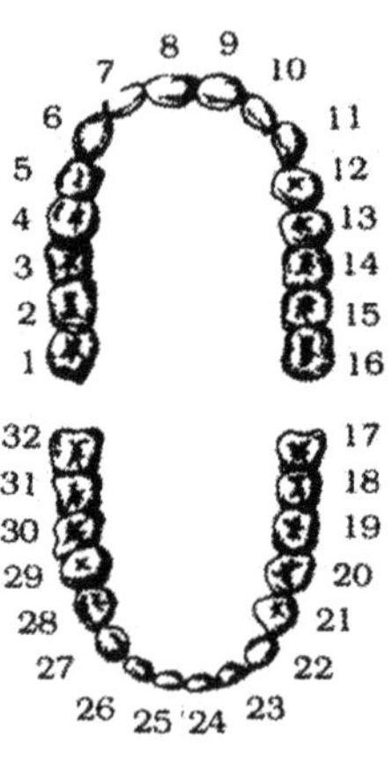

Human mandible in right lateral view

1-16-17-32 Third molars
2-15-18-31 Second Molars
3-14-19-30 First Molars
4-13-20-29 Second Bicuspids

5-12-21-28 First Bicuspids
6-11-22-27 Cuspids
7-10-23-26 Lateral Incisors
8-9-24-25 Central Incisors

against which the tongue presses during the formation of certain sounds. With cuspid canines and teeth with cusps due to the impalement of sounds upon exhalation, language is not possible. The reduction of the canines to the same size and height as the other teeth, along with the uniform, smooth, and even profile of the entire set of teeth, makes the passage of air without disturbances possible, and the quality of sounds is not affected (page 29).5

In reference to the canine teeth, Darwin says (page 424): *"In man, the canine teeth are perfectly efficient instruments of mastication, but the true character, as Owen* 6 *remarks, 'Is indicated by the conical form of the crown, which terminates in an obtuse point, is convex outward and flat or sub-concave within, at the base of which surface there is a feeble prominence. The conical form is best expressed in the Melanian races, especially the Australians. The canine is more deeply implanted, and by a stronger fang than the incisors.' Nevertheless, this tooth no longer serves man as a special weapon for tearing his enemies or prey; it may, therefore, as far as its proper function is concerned, be considered as rudimentary. In every large collection of human skulls, some may be found, as Häkel* 7 *observes, with the canine teeth projecting considerably beyond the others in the same manner as in anthropomorphous apes, but in a less degree. In these cases, open spaces between the teeth in one jaw are left for the reception of the canines of the opposite jaw. An inter-pace of this kind in a Kaffir skull, figured by Wagner, is surprisingly wide.* 8 *Considering how few are the ancient skulls which have been examined, compared to recent skulls, it is an interesting fact that in at least three cases the canines projected largely; and in the Naulette jaw they are spoken of as enormous."* 9

On page 425, Darwin comments: *"Of the anthropomorphous apes, the males alone have their canines fully developed; but in the female gorilla, and in less degree in the female orang, these teeth projected considerably beyond the others; therefore, the fact, of which I have been assumed, that women sometimes have considerably projecting canines is no serious objection to the belief that their occasional great development in man is a case of reversion to an ape-like progenitor. He who rejects with scorn the belief that the shape of his own canines and their occasional great development in other men are due to our early forefathers having been provided with these formidable weapons, will probably reveal, by sneering, the line of his descent. For though he no longer intends, nor has the power, to use these teeth as weapons, he will unconsciously retract his 'snaring muscles' (thus named by Sir C. Bell),* 10 *so as to expose them ready for action, like a dog prepared to fight."*

It is very possible that the rotation of the cranium, the reduction of the snout, the steady process of brain enlargement, the reduction of tooth size, and the disappearance of the diastema were all features that developed simultaneously, allowing the formation of the conditions for the creation of language.

The reduction of teeth to the shape and size of human teeth was instrumental in producing the different sounds that build words and compose spoken language. The conclusion here is that:

The teeth that we consider human are human because humans possess language. 11

Teeth are so important for the existence of language that they appear in a baby long before the baby starts to speak.

Considering this situation, the inference can be made that from the very moment when a tooth can be labeled as human or with human-like features, it belonged to a living being that possessed, or was on the way to possessing language.

Hence, language could be very old, since *australopithecines* have "human-like" teeth and therefore are considered hominids, and they are at least four to five million years old. Had they a language as developed as ours? Probably not; but a rudimentary form of the present language, probably.

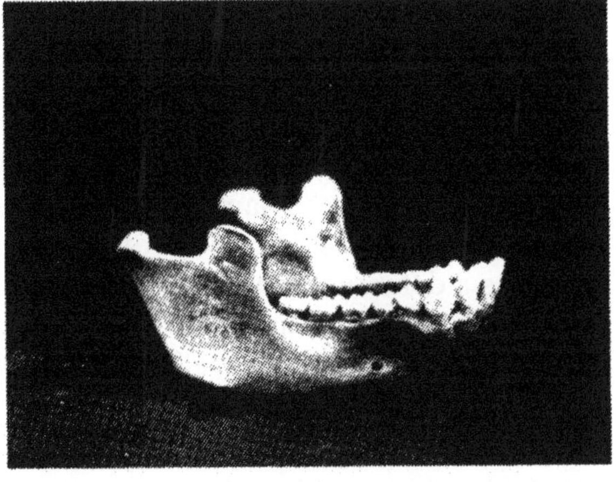

a

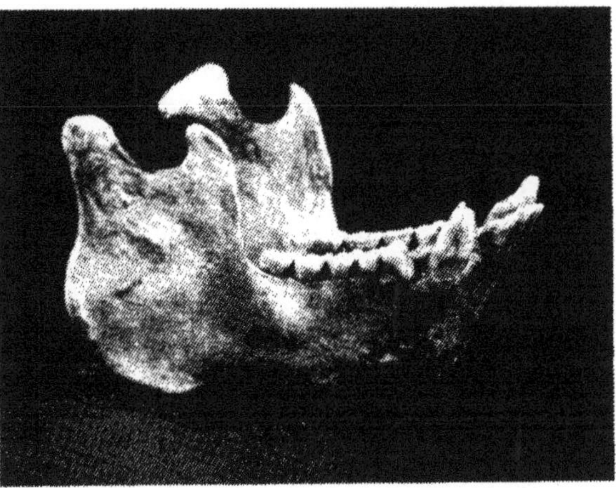

b

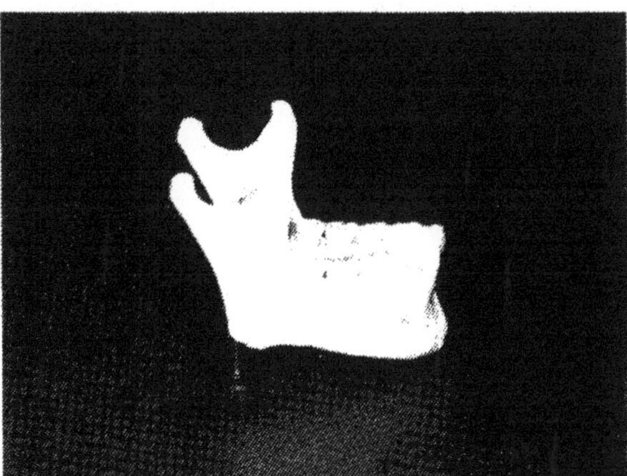

c

Mandible - lateral view
a) chimpanzee b) gorilla c) human

The appearance of anatomical details favoring the development of language

Vertical position of the body and rotation of the head with respect to the spinal column.

Rounded head allowing greater cranial cavity capacity and placement of the face in a near-vertical position.

A vertical face without prognathism allowing freedom of movement of the facial muscles essential for speech.

Enlarged cerebral capacity and enlargement of the various areas of the human brain, especially the areas associated with abstract thought and the development of speech.

The lack of enlarged canines and the lack of diastema.

The arched form of the dentition corresponding to the form of the tongue.

The appearance of vocal cords.

Canine teeth according to Darwin

When Darwin addressed the subject of the canine teeth, he compared gorilla and human canine teeth. He assumed that the fact that the male gorilla has canines much more developed than the female meant that the male gorilla used the canines as a weapon to defend the family group. Upon comparing the developed canines of the male gorilla and the insignificant canines of the human being, Darwin deduced that man lost his need for canines as a defense weapon long ago when he learned to use arms. Therefore, it is deduced that the existence of arms must date back to the oldest hominid dentition found with reduced canines. Darwin never suggested that the reduced canines found in human beings could be the product of another force. In the study here, the conclusion is reached that the reduced canines and characteristic arched shape of the mandible and the maxilla and uniformity of height and shape of the teeth are fundamental for the existence of language. We could be assured that human ancestors never had enlarged canines, and that the human line of evolution presented this selective dental characteristic in relation to the development of language.

Canine Teeth According to Darwin

When Darwin addressed the subject of canine teeth, he compared gorilla and human canine teeth. He assumed that because the male gorilla has canines much more developed than the female, the male gorilla used the canines as a weapon to defend the family group. Upon comparing the developed canines of the male gorilla and the insignificant canines of the human being, Darwin deduced that man lost his need for canines as a defense weapon long ago when he learned to use weapons. Therefore, he deduced that the existence of weapons must date back to the oldest hominid dentition found with reduced canines. Darwin says (page 435): *"The early male forefathers of man were, as previously stated, probably furnished with great canine teeth; thus as they gradually acquired the habit of using stones, clubs, or other weapons, for fighting with their enemies or rivals, they would use their jaws and teeth less and less. In this case, the jaws, together with the teeth, would become reduced in size, as we may feel almost sure from innumerable analogous cases."*

Darwin never suggested that the reduced canines found in human beings could be the product of another force. In this study, the conclusion reached is that **the reduced canines and characteristic arched shape of the mandible and the maxilla and uniformity of height and shape of the teeth are fundamental for the existence of language.** We could be assured that human ancestors never had diastema and enlarged canines. The human line of evolution presented this selective dental characteristic in relation to the development of language, simultaneously with the enlargement of the brain.

Notes

1. Dr. Webb, "Teeth in Man and the Anthropoid Apes," as quoted by Dr. C. Carter Blake in *Anthropological Review*, July 1867, p. 299.
2. Owen, *Anatomy of Vertebrates*, vol.iii, pp. 320 and 325.
3. "On the Primitive Form of the Skull", English translation in *Anthropological Review*, Oct. 1868, p. 426.
4. Professor Montegazza writes to mefrom Florence that he has lately been studying the last molar teeth in the different races of man, and has come to the same conclusion as that given in my text, viz., that in the higher or civilized races, they are on the road toward atrophy or elimination.
5. *Man, the Paradoxical Primate*, Jordi Fuentes, 1985.
6. *Anatomy of Vertebrates*, vol. iii, 1868, p. 323.
7. *Generelle Morphologie*, 1866, B. ii, s., clv.
8. Carl Vogt's *Lectures on Man*, English translation, 1864, p. 151.
9. C. Carter Blake, on a jaw from La Naulette, *Anthropological Review*, 1867, p. 295; Achaaffhausen, ibid., 1868, p. 426.
10. *The Anatomy of Expression*, 1844, pp. 110 and 131.
11. *Man, the Paradoxical Primate*, Jordi Fuentes, 1985.

XIV

The Human Face

The human face is the most expressive within the animal world (page 32). 1

Many animals use body language to communicate with their kin, but the facial muscles play only a reduced part in this communication, as the tail, body posture, etc. are more expressive and important in order to communicate their feelings. These feelings could be of superiority, inferiority, submission, equality, and so forth.

The many different expressions of the human face are possible due to the quantity and mobility of the frontal facial muscles, especially those muscles that form the mouth cavity, lips, and cheeks. These muscles are directly related to the possibility of spoken or phonetic language. The infinite number of variations produced by the different lip positions help to modulate and create varieties of sound. The sounds created upon the vibration of vocal cords are transmitted with the emission of air passing through the mouth cavity, undergoing modifications according to the position of the tongue with respect to the palate and teeth.

The form of the mouth cavity is modified by the cheek muscles and especially by the muscles that make up the lips. The intensity and volume of sounds are modified according to the position that these muscles take i.e., circular shape, funnel shape, closed, open, and likewise. Apart from these direct variations of the kind and quality of the sound produced by the facial muscles, other facial muscles not directly involved in producing language aid spoken language with facial expression. Facial expression helps to reveal the true intention behind the spoken word. There are many different expressions that use only the facial muscles to communicate our feelings or intentions, like the raising of an eyebrow, the tongue in the cheek, the kniting of the eyebrows, and the funneling of the lips. Facial muscles are those muscles that allow us to smile, laugh, cry, or express sadness or indifference. All of these expressions constitute body language and serve to communicate to another the state of animation or intention of he who is expressing, revealing hidden emotions as he speaks (page 32).2

Facial muscles make man the only primate that can sing and whistle.

Notes

1. *Man the Paradoxical Primate*, Jordi Fuentes, 1985.
2. Ibid.

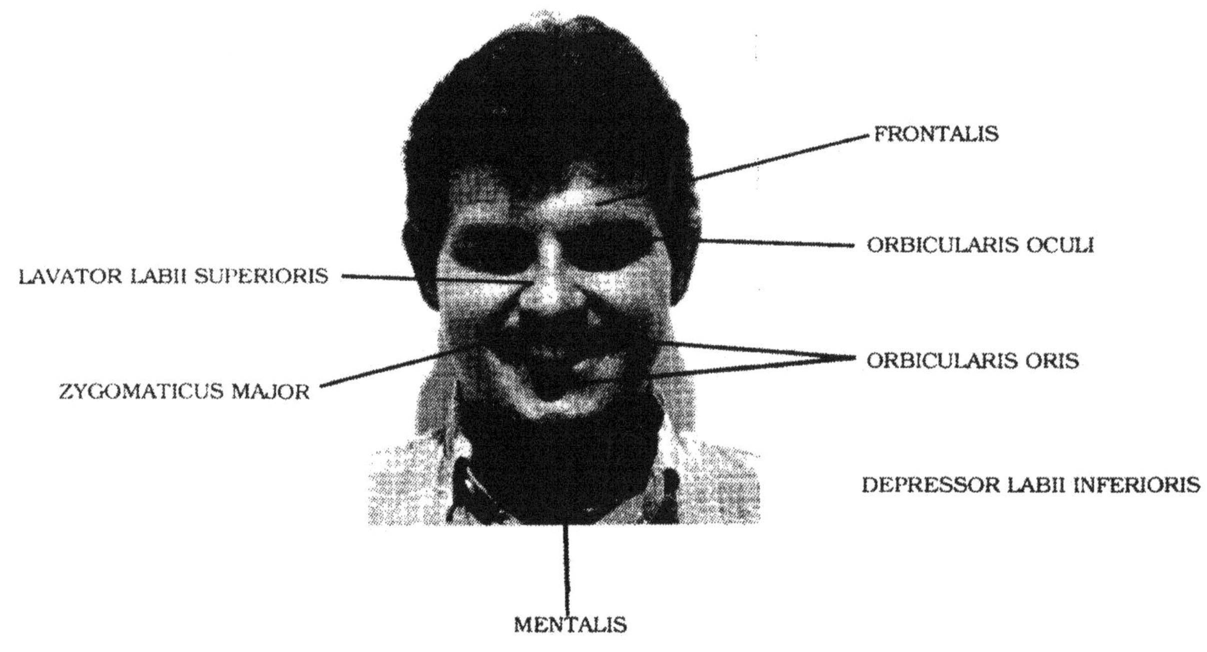

Muscles of facial expression
Surface anatomy photograph of the anterior facial muscles.

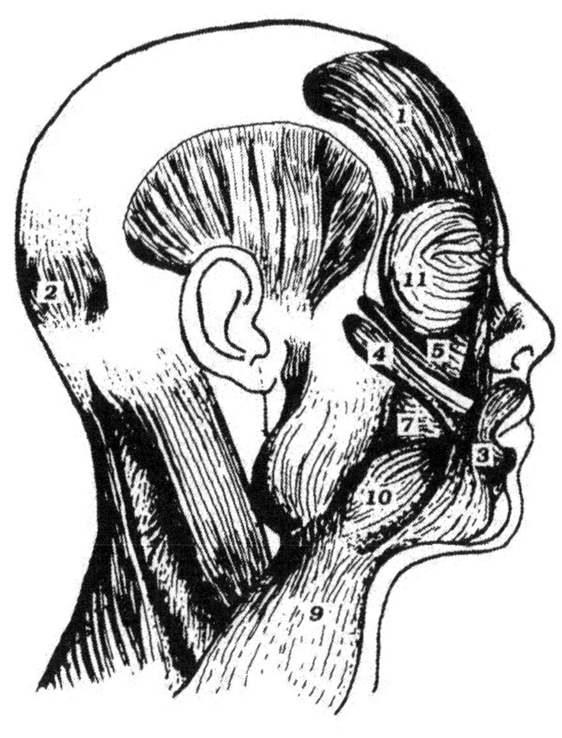

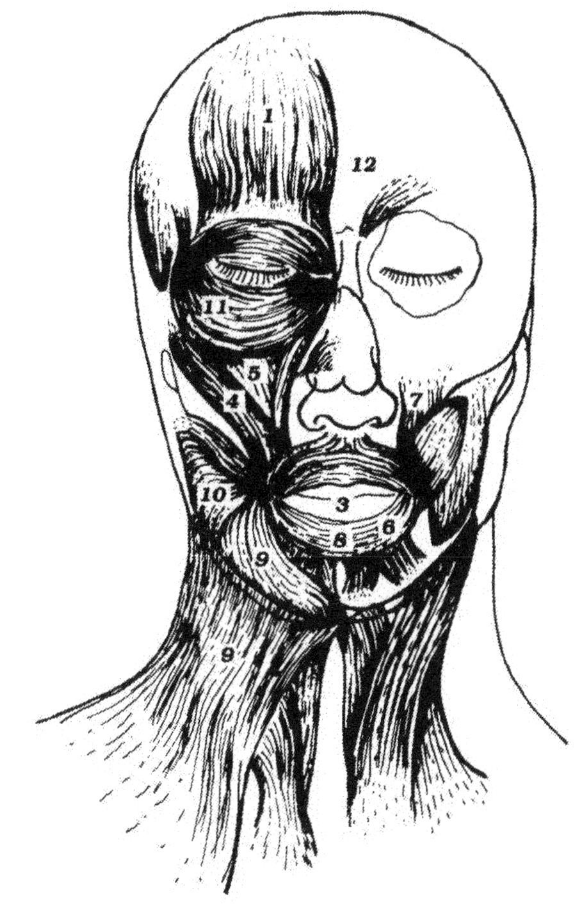

Muscles of facial expression

MUSCLE	ACTION
1. Frontalis	*Raise eyebrows, draws scalp forward and wrinkles forehead horizontally.*
2. Occipitalis	*Draws scalp backward.*
3. Orbicularis Oris	*Closes lips, shapes lips during speech, compresses lips against teeth.*
4. Zygomaticus Major	*Draws angle of mouth upward and outward as in laughing or smiling.*
5. Levator Labii Superioris	*Elevates upper lip.*
6. Depressor Labii Inferioris	*Depresses lower lip.*
7. Buccinator	*Major cheek muscle; compresses cheek as in blowing air out of the mouth Causes the cheeks to cave in as in sucking.*
8. Mentalis	*Protrudes and elevates lower lip and pulls skin of the chin up as in pouting.*
9. Platysma	*Draws outer part of lower lip downward and backward as in pouting; depresses mandible.*
10. Risorius	*Draws angle of the mouth laterally as in tenseness.*
11. Orbicularis Oculi	*Closes eye.*
12. Corrugator Supercilii	*Draws eyebrow downward as in frowning.*
13. Levator Palpebrae Superioris	*Elevates the upper eyelid (not shown).*

XV

Man's Hand

Man's hand has really evolved little within the animal kingdom. This statement may sound incredible until we compare man to other animals. Basically, the structure of man's hand does not differ greatly from that of the amphibian foot (the amphibian is considered the first animal on Earth to be able to breathe air). He possesses the same five digits that in relation to the palm, cannot be considered elongated. When I say that man's hand is very little evolved, I do not mean that it lacks specialization. Man's hand has achieved great specialization without departing from a rudimentary basic structure (page 34).1

With reference to this subject, Darwin makes this statement (page 426): *"The muscles of the hands and arms parts which are so eminently characteristic of man are extremely liable to vary, so as to resemble the corresponding muscles in the lower animals."* If they are so characteristic of man, how can they, at the same time, resemble those of the lower animals? Frankly, my limited mental capacity does not let me see the idea behind this statement; if they resemble each other, even if the size or shape is different, they are not so "eminently characteristic of man." Darwin continues his exposition (page 433): *"The structure of the hand in this respect may be compared with that of vocal organs, which in the apes are used for uttering various signal-cries, or, as in one genus, musical cadences; but in man closely similar vocal organs have become adapted through the inherited effects of use for the utterance of articulate language."* Where is the comparison?

Darwin continues (page 433): *"Turning now to the nearest allies of men, and therefore to the best representatives of our early progenitors, we find that the hands of the Quadrumana are constructed on the same general pattern as our own, but are far less perfectly adaptable for diversified uses. Their hands do not serve for locomotion so well as the feet of a dog, as may be seen in such monkeys as the chimpanzee and orang* [here Darwin calls chimpanzees and orangs, "monkeys"] *which walk on the outer margins of the palms, or on knuckles. Their hands, however, are admirably adapted for climbing trees. Monkeys seize the branches or ropes, with the thumb on one side and the fingers and palm on the other, in the same manner as we do. They can thus also lift rather large objects, such as the neck of a bottle, to their mouths. Baboons turn over stones, and scratch up roots with their hands. They seize nuts, insects, or other small objects with the thumb in opposition to the fingers, and no doubt they thus extract eggs and young from nests of birds. American monkeys beat the wild oranges on the branches until the rind is cracked, and then tear it off with the fingers of the two hands. In a wild state they break open hard fruits with stones. Other monkeys open mussel-shells with the two thumbs. With their fingers they pull out thorns and butts, and hunt for each other's parasites. They roll down stones, or throw them at their enemies. Nevertheless, they are clumsy in these various actions, and, as I have myself seen, are quite unable to throw a stone with precision. It seems to me far from true that because 'objects are grasped clumsily" by monkeys,' a much less specialized organ of prehension" would have served them equally well with their present hands. On the contrary; I see no reason to doubt that more perfectly constructed hands would have been an advantage to them, provided that they were not thus rendered less fitted for climbing trees. We many suspect that a hand as perfect as that of man would have been disadvantageous for climbing ; for the*

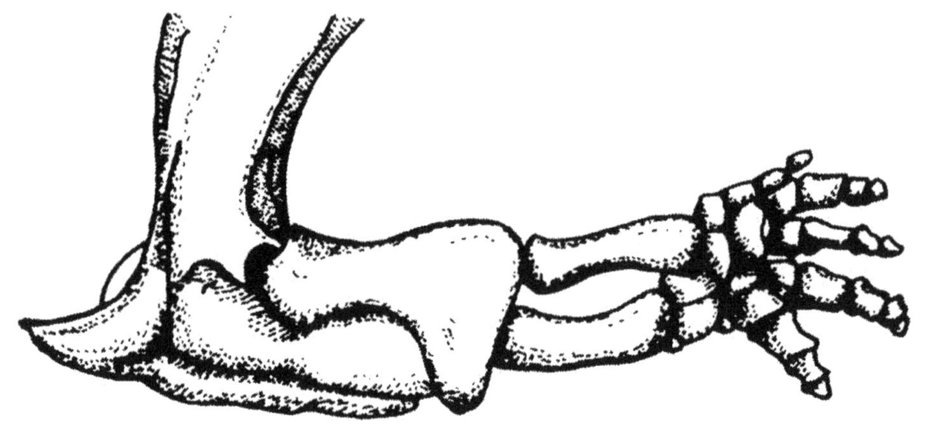

Early amphibian limb

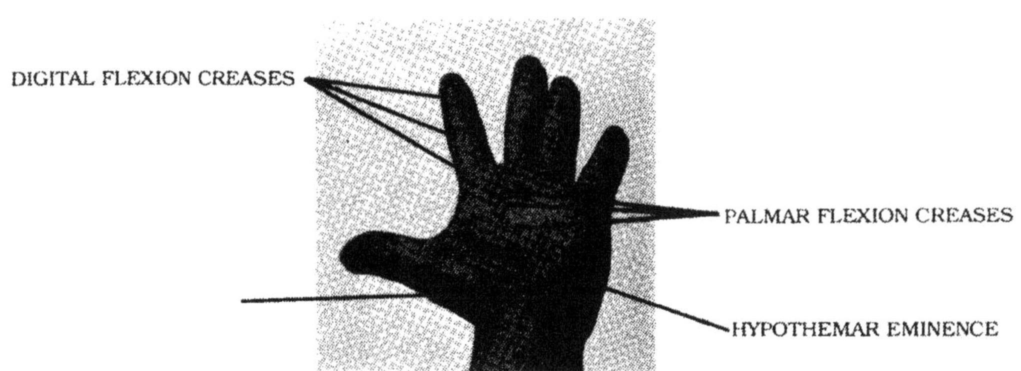

Palmar Surface

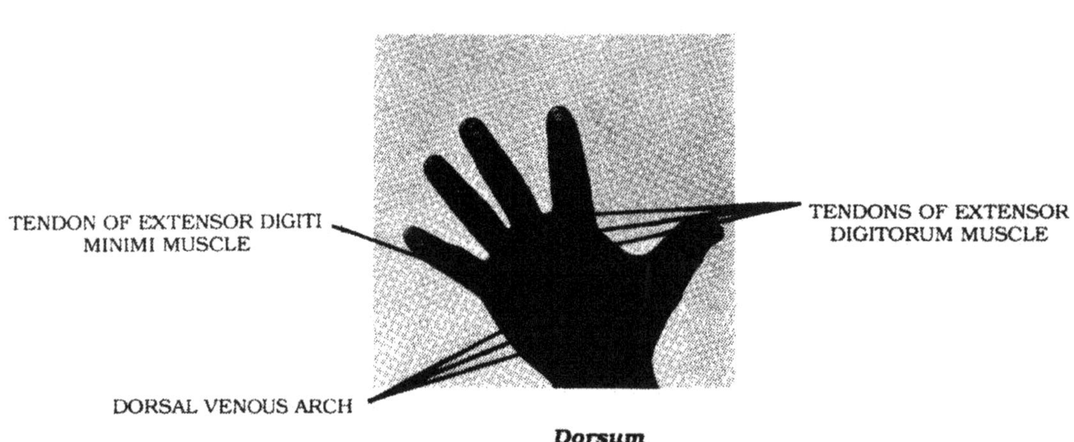

Dorsum

Human hand
Surface anatomy photograph

Human hand
Plier-like movement allowing the fingers and opposable thumb to pick up small objects.

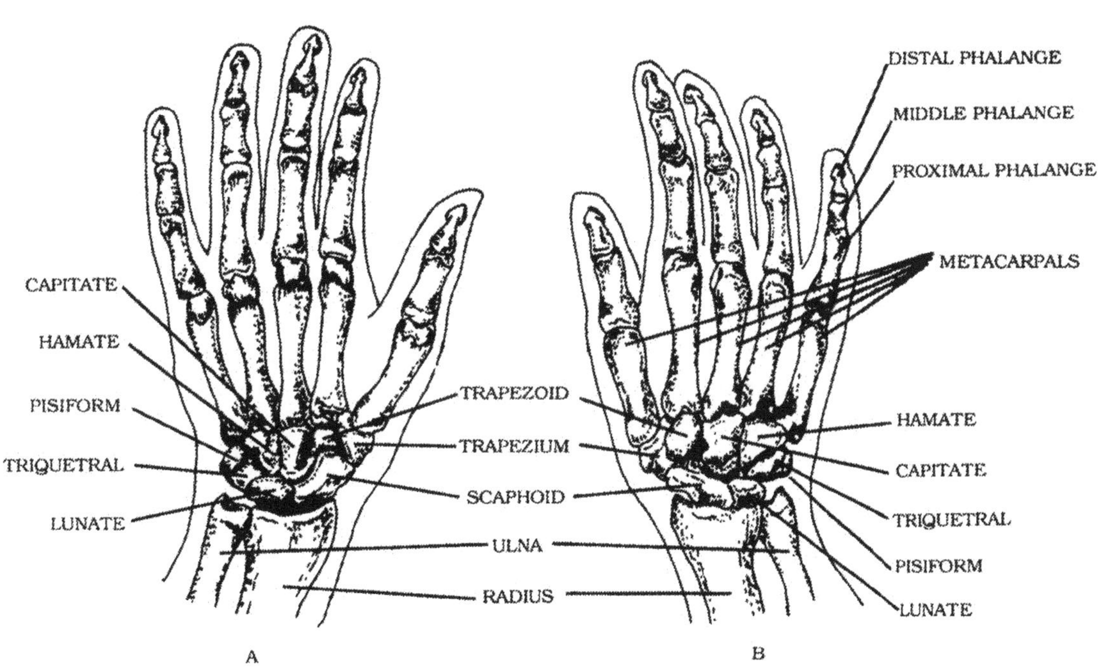

Human right wrist and hand. *a) anterior view b) posterior view*

most arboreal monkeys in the world, namely, Ateles in America, Colobus in Africa, and Hylobates in Asia, are either thumbless or their toes are partially cohered, so that their limbs are converted into mere grasping hooks." 2

A detailed study of man's hand takes us to the situation to affirm the following:

The changes that have taken place in man's hand are not spectacular, but very positive.

All primate hands (except for the more primitive Platyrrhines) have nails instead of claws. Primate hands are elongated and tend to have conical-shaped fingers. Man's fingers are more cylindrical in shape and possess acute sensibility, due to the softer tissue found in the fingertips, with many sensory receptors.

It has been considered of great advantage that man has an opposable thumb (pollex), allowing plier-like usage of the hands. Some consider this ability the key to why man has achieved so much versatility, which aided our progress as technological animals. Undoubtedly our opposable thumb is very important, as anyone knows who has injured this thumb and then tried to pick up anything, especially small things, without the use of his thumb. However, all of the successfulness of man's hand should not be solely attributable to our thumb; just as important is the ability of our fingertips to transmit information upon touching an object (see the somesthetic association area in *"Sensory Areas of the Cerebrum"*). In this case, our four fingertips appear to be more sensitive than our thumb. If you ask a friend to guess what an object is without looking at it, notice the method he uses to obtain the answer. He will spread his hand and run his hand over the object so that his fingertips have maximum contact with the object. He will most likely leave the thumb out of the entire process. The reason for this is that our thumbs are not in the same flat position as our fingers. If you study your hand you will notice that our thumb is slightly rotated inward; therefore, when someone is touching something, it is harder for the tip of his thumb to become involved. Observe a blind person reading by the Braille method (page34).3

Both qualities, that of an opposable thumb and that of having acute sensibility in the fingertips, make man's hand an extraordinary instrument. This instrument, controlled by man's highly developed brain, is the tool by which present human society has been constructed in dominance over all other life forms (page 34).4

Philosophers and anthropologists have considered the human hand as the base or foundation on which civilization was initiated and evolved to the advanced technological point at which we are at the end of the twentieth century. This conclusion was reached as a consequence of relating the very specialized human hand and the product that, with this tool, man has achieved. The existence of the hand was initiated long before it could be used and helped by the parallel development of the brain. The real cause of the existence of our civilization is not the hand itself, but the little developed human newborn who forced the creation of the specialized hand, and, from there, the continuous ascension to the place humans occupy today.

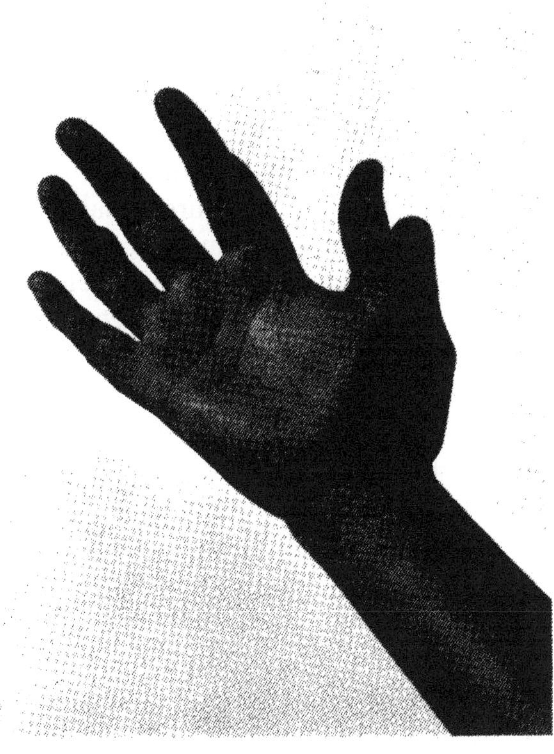

a

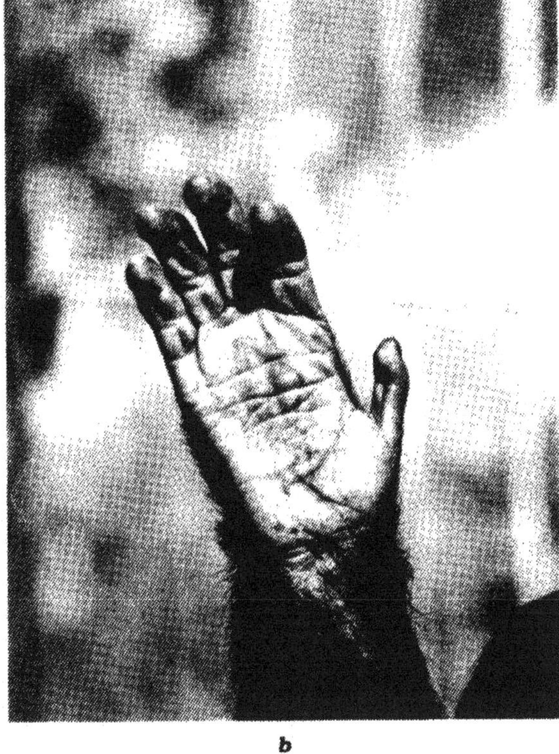

b

c

Thumb placement

a) Human hand demonstrating not only a well defined thumb, but also a thumb which is opposable and well developed. This thumb can be used in plier-like functions. b) Chimpanzee hand. The thumb is differentiated from the fingers but it is only partially opposable. The size of the chimpanzee thumb indicates that it is not functional and cannot be used in plier-like functions. Its position on the hand with relation to the fingers is much further removed than its counterpart on the human hand. The chimpanzee thumb's small size, its inability to function well upon picking up an object and its low position on the hand would all be indications that this thumb will eventually disappear. This suggests that the chimpanzee hand could have evolved from a hand similar to a human hand, possibly from a terrestrial primate, a common ancestor to both man and chimpanzee. Upon specializing in arboreal life the hand adapted by becoming more of a hook than a plier to enable the chimpanzee to have a better grip on branches. Therefore, the thumb no longer is required, nor does it offer any advantage and would tend to disappear, such as occured in the genus Ateles c) whose hand only has four fingers. Their thumb possibly disappeared as it became a disadvantage for arboreal life. In other primates the thumb exists but is not well defined or not defined at all.

Two-Phalanged Thumbs

The thumbs of the hand, as well as their equivalent on the feet, are notable in their difference from the other fingers or toes, in that they possess only two phalanges instead of three. There are those who have theorized that, in reality, the thumb and the big toe have three phalanges, and what has actually happened is that the corresponding metacarpus and metatarsus have disappeared, and the first phalanges have replaced these. This theory is not very clear, since there does not appear to be enough evidence to support the theory that the metacarpus and metatarsus are those which have disappeared. In the case of the big toe, it would be advantageous to have only two phalanges in place of three, since the big toe receives the full weight of the body when, in the act of walking, the foot separates from the ground. Another phalange would be disadvantageous since it would lengthen the big toe's "shove off" off the ground, therefore making walking more difficult and less efficient.

Notes

1. Man, the Paradoxical Primate, Jordi Fuentes, 1985.
2. In *Hylobates syndactylus*, as the name expresses, two of the toes regularly cohere; and this, as Mr. Blyth informs me, is occasionally the case with the toes of *H. agilis*, and *leuciscus*. Colobus is strictly arboreal and extraordinarily active (Brehem, *Thietleben*, B.i.S. 50), but whether a better climber than the species of allied general, I do not know. It deserves notice that the feet of the sloths, the most arboreal animals in the world, are wonderfully hook-like.
3. *Man, the Paradoxical Primate*, Jordi Fuentes, 1985.
4. Ibid.
5. Ibid.

Thumb placement

a) Human hand demonstrating not only a well defined thumb which is opposable and well developed. This thumb can be used in plier-like functions. b) Chimpanzee hand. The thumb partially opposable. The size of the chimpanzee thumb indicates that it is not functional and cannot be used in plier-like functions. Its position in the hand with relation to the fingers is much further removed than its counterpart on the human hand. the chimpanzee thumb's small size, its inability to function well upon picking up an object and its low position on the hand would all be indications that this thumb will eventually disappear. This suggests that the chimpanzee hand could have evolved from a hand similar to a human hand, possibly from a terrestrial primate, a common ancestor to both man and chimpanzee. Upon specializing in arboreal life the hand adapted to this function by becoming more of a hook than a plier to enable the chimpanzee to have a better grip on branches. Therefore, the thumb no longer is required, nor does it offer any advantage and would tend to disappear, such as occurred in the genus *Ateles*, c) whose hand only has four fingers. Their thumb possibly disappeared as it became a disadvantage for arboreal life. In other primates the thumb exists but is not well define or not defined at all. ('Man the Paradoxical Primate, Jordi Fuentes, 1985).

XVI

Stereoscopic Vision

Much has been written about a likely insectivore ancestor giving rise to primate vision. This hypothesis reflects our present knowledge, as the oldest mammalian fossils have been insectivores, as evidenced by their dentition, form, size, and resemblance. The information regarding the first mammals is incomplete however and the fact that other fossils with mammalian characteristics during this same period have not been found yet does not mean that they never will be found. Perhaps one day other mammalian fossils will be found and it will be necessary to completely change our ideas about the mammals from which we descend (page 45).1

Presently, the possibility that some animals still considered dinosaurs due to their pelvical structure could have actually been warm-blooded (a step closer to mammals is being discussed). It is possible that they were mammals that conserved reptilian characteristics (page 45), 2 but so far it looks like mammals evolved from another branch in the animal genealogical tree.

It has been speculated that, little by little, our primate ancestors developed frontal vision, with the eyes rotating from the side of the head too the front of the face. Frontal placement of the eyes has been the basis for the idea that stereoscopic vision is possible, and hence the ability to measure distance and depth so necessary for arboreal life. This idea had been repeated without reflection about it by anyone writing on the subject of evolution. Authors copied each other, and when they found an idea that they considered "brilliant," they absorbed it and repeated it in any work they produced without having reflected on it. With reference to this idea on the origin of vision, perhaps if anthropology and biology had worked closer together, a more probable theory would have been developed (page 45).3

With reference to frontal placement of the eyes being a requirement for stereoscopic vision, one would have to deduce that, therefore, animals whose eyes are placed laterally would not have stereoscopic vision, and therefore are not capable of judging depths and distances. Due to my interest in biology, I have spent many years observing animals, and I will state that I don't know any animal (including the majority of whales, as it hasn't been proved with all whales) that does not have frontal vision regardless of the placement of the eyes. Frontal vision is a characteristic shared by all vertebrate animals, including mammals, birds, amphibians, reptiles, and fish (page 45),4 excluding logically these subterranean mammals that live underground making the eyes unnecessary; their eyes have been atrophied.

What is absolutely true is that: **All animals can calculate depth and distance.** Notice a horse running an obstacle course, and you can appreciate its ability to make precision distance calculations concerning the obstacle and the distance between itself and the obstacle, so that it misses it. The same can be said for arboreal animals, such as the squirrel, with lateral placement of their eyes. Squirrels are superb judges of distance and depth (page 45).5

There are many species of animals that have a property not shared by humans that permits them not only to calculate depths and distances, but also to correct the refraction of light between air and water and vice versa. All birds that fish from the air (here, diving birds are excluded) can correct this refraction and aim accurately at their prey under water while standing still or while flying. This same property is found also in marine mammals like dolphins and whales, who not only can calculate depth and distance, but can correct the refraction of light between water and air. This is the reason that in all aquariums that exhibit dolphins and orcas, or killer whales, these animals can spring out from the water and catch an object suspended high over the water, a toy or a piece of food.

Conclusion: Stereoscopic vision is not the fruit of the frontal placement of eyes. All animals possess frontal and stereoscopic vision regardless of the placement of their eyes, and therefore to attribute stereoscopic vision as a special characteristic of primates is an error that must be discarded.

Notes

1. *Man the Pradoxical Primate*, Jordi Fuentes, 1985.
2. Ibid.
3. Ibid.
4. Ibid.
5. Ibid.

XVII

Man's Specialization

If we look at the physical appearance of man, we could come to the conclusion that man specializes in not having a specialization; in other words, as Dr. Niles Eldredge says, "We are, in short, more of a hybrid of generalist/specialist than most other creatures appear to be." 1 His success as a species would be due to the fact that he can eat a great variety of foods and live in opposite climates under conditions diametrically opposed. These are valid judgments if we only consider man's physical body as a living machine. Is this living machine, in reality, as little specialized as it appears? Of course, the answer is no; and, in fact, although man appears to be little specialized, he is extremely specialized (page 46).2 We will see that in this chapter.

With regard to this apparent unspecialized condition of the human being, Darwin states (page 443 and 444): *The Duke of Argyll, for instance, insists that '(The human frame has diverged from the structure of brutes, in the direction of greater physical helplessness and weakness. That is to say, it is a divergence which of all others it is most possible to ascribe to mere natural selection.) He adduces the naked and unprotected state of the body, the absence of great teeth or claws for defense, the small strength and speed of man, and his slight power of discovering food or of avoiding danger by smell. To these deficiencies there might be added one still more serious, namely, that he cannot climb quickly, and so escape from enemies. The loss of hair would not have been a great injury to the inhabitants of a warm country. For we know that the unclothed Fuegians can exist under wretched climate. When we compare the defenseless state of the man with that of the apes, we must remember that the great canine teeth with which the latter are provided are possessed in their full development by the males alone, and are chiefly used by them for fighting with their rivals; yet the females, which are not thus provided, manage to survive."*

In the case of human beings, evolution began favoring those traits enhancing cerebral activity, and over the ages, the mind played an increasingly important role in man's evolution.

Cerebralization is man's real specialization.

From cerebralization many other traits that separate man from other animals are derived. Dependent on cerebral activity are such outstanding human characteristics as language, and, with language, the ability to transmit technology and, later, transmit ideas and abstract concepts, with which man organizes his life and studies, and tries to understand the world that surrounds him.

There are various animals that have developed social living, or living in groups composed of individuals who cooperate in certain activities. In this way, they have developed a stratification of hierarchies, with the rewards of living in harmony under the dominance of one animal. Human beings have developed social living with hierarchical stratification; this organization has been facilitated by the human ability to communicate by language and, therefore, as mentioned earlier, to transmit technology and abstract ideas (page 46).3 In the rest of the animal kingdom, the lack of language is not absolute, but limited and complemented mostly by communication through body language and a limited vocalization for some species.

In other groups of animals in which a hierarchy exists, dominated by one member of the group, the dominant member demands subjugation by the rest of the group due to his physical superiority. The group both recognizes and accepts this dominance. When this dominant animal becomes too old, he is replaced usually by a younger, stronger animal who is capable of challenging the dominant animal and suppressing any other rivals by the use of physical force and strength.

It is very possible that, at one point of our evolution, this was also the human way of establishing dominance or leaders; he who had the greatest physical strength ruled the group. However, undoubtedly the time came when abstract ideas and their strength were also considered. Only the unique development of the human brain has allowed the capacity of producing and storing abstract concepts foreign to physical reality. He who was most able to express himself and express ideas about the future and the present, even though his physical strength may not have been great, would acquire greater dominion over the group through his spiritual appeal to its members. He became the spiritual leader of the community thus the origin of the priest, healer, sorcerer, and likewise. At times, both physical strength and spiritual dominance could be present in the same individual, but more than likely these characteristics were present in two different individuals, hence, the king and the priest, the leader and the sorcerer, the church and the state (page 46).4

Man's cerebral specialization is directed toward abstract ideas applied to better the life and survival of the group, and, unfortunately, as sometimes is witnessed in our society, to the destruction of the social group (page 46).5

At the same time, this cerebral specialization is due to the forces created by a helpless infant obligating bipedalism and the necessity to have a nest which meant that the majority of the activities would be organized around the nest or place of residence. Little by little, man's ancestors would have time and possibilities to experiment with food and materials, creating better living conditions. Simultaneously, the mind was forced to play an increasingly important role in the life of the individual, as much in the receipt of information from the world that surrounded him as in the ability to reflect upon this information in order to judge situations and results of activities (page 46, 47).6

As the brain grew in size, the ability to accumulate and use stored information grew. This information file was first composed of personal experiences and, later, through language, of vicarious experiences (page 47).7

It appears as though nature is looking for equilibrium between physical specialization and mental specialization. When an animal develops his body toward physical specialization, the capability to develop his brain diminishes, or, better, does not increase in the same proportion, and vice versa. An example of this would be the chimpanzee and the gorilla, whose ancestors (probably also man's ancestors) began to develop intellectual capacity until reaching approximately 380 grams of brain weight. At this point, the chimpanzee and gorilla's specialization became their bodies' occupation of a convenient ecological niche, and the development of cerebral capacity did not progress. That is to say, their bodies began to specialize in characteristics to facilitate arboreal life and their cerebral development remained the same. The increase in weight of the gorilla brain is only a reflection of an animal that is larger than the chimpanzee, but the size of the brain is proportionally the same (page 47).8

On the other hand, the human ancestor never stopped developing his cranial capacity until he became "humanized." In order to survive as a species without a physical speciality, he had to develop a mental speciality. In augmenting his brain capacity, his chances for survival also augmented in the search for food and in securing protection for himself and the next of kin (page 47).9

Man is physically weak in comparison with other mammals of his size, nevertheless, with the aid of his brain, he has been able to overcome animals many times stronger and larger than he, too become the *King of Creation*, as we like to call ourselves. We only hope that man will continue to use his intelligence for self-perpetuation and not for self-destruction (page 47).10 That future is not clear.

Notes

1. *Time Frames* by Niles Elredge, Simon & Shuster, N.Y., 1985, p. 179.
2. *Man the Paradoxical Primate*, Jordi Fuentes, 1985.
3. Ibid.
4. Ibid.
5. Ibid.
6. Ibid.
7. Ibid.
8. Ibid.
9. Ibid.
10. Ibid.

XVIII

Power of Concentration

In discussing the cerebralization of man as the fundamental difference between man and other animals, one ability that makes the fact that man can store and use an enormous amount of information at will meaningful has, for the most part, been overlooked. This ability is the power of concentration (page 48).1

In this work, I use the term "concentration," which is defined by the *Webster's New Universal Unabridged Dictionary* as: *"fixed or close attention."* The verb "concentration" is defined as *"To direct one's thoughts or effort; to fix one's attention (on or upon)."* Darwin uses the word "attention," which is defined by the same dictionary as "the ability to give heed or observe carefully." At first sight, one can take both words to be synonymous, so their use does not make a major difference. But I think that concentration requires a longer period of thought in order "to fix attention." One can pay attention to a situation and, after that, not ponder any more about it. On the other hand, concentration means that the attention of the individual to a given fact, idea, etc. is fixed, and persists longer in the mind of the individual sometimes for days, weeks, months, and even years. Thinking abut this particular interest will continue until the individual solves, at least in his mind, the problem that is the subject of his preoccupation.

Regarding "attention," Darwin says (page 452): *"Hardly any faculty is more important for the intellectual progress of man than attention. Animals clearly manifest this power, as when a cat watches by a hole and prepares to spring on its prey. Wild animals sometimes become so absorbed when thus engaged that they may be easily approached."*

Animals have the ability to reason imprinted in the characteristics of the species. This is the reason we can observe two animals from the same species reacting very similarly in like circumstances. You could say that all the members of certain species of animals follow an established pattern of behavior. This pattern of behavior is called instinct (page 48):2

If we begin to study a given animal more closely with respect to instinct, we will see that its behavior is not as rigid as it appears to be. Each individual animal will react slightly differently from another, and at a slightly different time. Even though one could say that these animals have instinct as a common denominator, each individual has a different personality and disposition that affect and determine its behavior. These differences in behavior are due to many things, the most important of which are: hereditary factors contained in its genes, individual experiences remembered, and the special conditions that surround it. Special stimuli, such as tension, hunger, thirst, and reproductive drive, cause similar but not identical reactions from an animal species (page 48).3

Human beings are not totally foreign to the instinctual behaviors mentioned previously, and also share many of the same characteristics as other animals. However, the one characteristic that makes up the fundamental difference between man and all other animals is his ability to concentrate. This ability to focus on an idea may last from a few seconds in some cases, to minutes and hours, to months, and even years. The ability to concentrate on an idea makes it possible for the brain to use all stored information relative to the idea and find the most appropriate solution to whatever the problem; hence, the ability to reflect (page 48).4

All animals, including man, are capable of receiving information through their senses and acting accordingly. The majority of animals react only according to these external stimuli, and are in a state of constant alertness, ready for any indication of external stimuli. The fundamental function of these animals' brains is to receive these stimuli and give the proper signals to the appropriate parts of the body to act (page 48).5

Darwin says (page 448): *"The principle of imitation is strong in man . . . No doubt, as Mr. Wallace has argued,* 6 *much of the intelligent work done by man is due to imitation and not to reason; but there is this great difference between the actions and many of those performed by the lower animals namely, that man cannot, on his first trial, make, for instance, a stone hatchet or a canoe, through his power of imitation. He has to learn his work by practice; a beaver, on the other hand, can make its dam or canal, and a bird its nest, as well, or nearly as well, and a spider its wonderful web, quite as well, the first time it tries as when old and experienced."*

Man also has these abilities, but he is unique in the animal kingdom in his ability to distract himself from immediate environmental stimuli and orient his thinking toward the development or reflection of one idea or the solution of one problem for whatever the period of time necessary. Darwin makes the following comment (page 452): *"The imagination is one of the highest prerogatives of man. By this faculty he unites former images and ideas, independently of the will, and thus creates brilliant and novel results."*

I partly agree with this comment, and I will separate two ideas: the positive, with which I agree, and, later, the one with which I do not agree. Imagination is the power to create mental images of what is not actually present. This act of creating images of what has never been experienced, or the possibility to combine previous experiences to create new images or ideas, is what we call "creative power." In other words, imagination is the power of the mind to separate its conceptions, recombining the elements as it pleases. The faculty of imagination is the basis for understanding and appreciating the works of others, as is the case with literature or art. Without imagination, we cannot develop any idea. This book could not have been written without the intervention of imagination. Without separating the conceptions held in my brain and recombining all the elements to create a different point of view and, with it, new possibilities, this book would be empty of interest, its pages would be a string of blanks.

Some animals create abstract concepts, as seen in the demarcation or marking out of boundaries, and the defense of this piece of territory in which the established boundaries are immaterial. That means that the animal possesses some sort of abstract idea or imagination about what constitutes its estate. If the owner of this imaginary territory discovers a trespasser, it pursues this intruder until it is expelled from the imaginary demarcation. That proves that imagination exists in animals, and is not exclusively characteristic of man. What is certain is that in man, imagination has reached a high state of development. I do not agree that imagination is a prerogative of man. "Prerogative" is defined as an exclusive right or privilege."

Some animals exhibit the ability to solve problems, but only when the solution to the problem can be found immediately, without demanding much time or attention. Power of concentration is very scarce in the zoological scale, and where it is found at all, it is extremely limited (page 48).7 Chimpanzees are known to solve problems that require imagination, and they show a certain degree of it. That means that they can process past experiences and extract from them the elements to find a solution to a problem they are facing. Their imagination is limited to the physical world around them, but as far as we know, given the limited information, we have, or, in other words, because of our ignorance, they are not able to process abstract concepts.

Regarding this point of view, Darwin says (page 451): *"It has been asserted that man alone is capable of progressive improvement; that he alone makes use of tools or fire, domesticates other animals, or possesses property; that no animal has the power of abstraction, or of forming general concepts, is self-conscious and comprehends itself; that no animal employs language; that man alone has a sense of beauty, is liable of caprice, has the feeling of gratitude, mystery, etc.; believes in God, or is endowed with a conscience. I will hazard a few remarks on the more important and interesting of these points. Archbishop Summer formerly maintained that man alone is capable of progressive improvement.* 8 *That he is capable of incomparably greater and more rapid improvement than is any other animal admits of no dispute; and this is mainly due to his power of speaking and handing down his acquired knowledge."*

I cannot accept the immediately foregoing paragraph a 100 percent, as many animals have a language, even if it is very different from ours. The fact that we do not understand their vocal communication does not invalidate their existence. I don't understand the hundreds of languages that the human species uses, but they exist despite my ignorance. About the *"greater and more rapid improvement"*..... or destruction (?) I have to admit that I am a cynic; so far, all of the technical improvements that I have been able to appreciate are pushing the human species to self-destruction and the destruction of the planet on which we live. "Improvement" is a word whose value is contingent upon a subjective judgment.

I have another disagreement with Darwin's paragraph, because man is not the only animal that "domesticated other animals." Vertebrate animals do not show this characteristic of domesticating other species for their benefit, but ants do. Ants domesticate and care for animals (insects), which they exploit to their benefit. They domesticate aphids (*Aphis*) and "scales", [We call "scales" any of a number of related sucking insects that attack plants] the same way that we do with cows, and they "milk" them. Some species of ants appear to be very active in practicing slavery. They raid other ant's nests and steal the larvae, taking them to their nest where they take care of them until they reach adult status, and then they are put to work for the benefit of the robbers.

Intelligence can be defined as the ability to solve a problem. Next to man, primates are those land animals that demonstrate the most ability in the solving of new problems (page48).9 Surprisingly one animal considered a "lower animal," the common rat (*Ratus norvegicus*), has been proved in laboratory experiments to have a very high power of reflection and the ability to solve different situations that we consider exclusive of "higher animals." Perhaps what happens is that we are ignorant of the mental capacity of many other species, because we did not have the opportunity to work with them as much as we did with laboratory rats.

In this respect, Darwin states (page 453): *"Of all the faculties of the human mind, it will, I presume, be admitted that reason stands at the summit. Only a few persons now dispute that animals possess some power of reasoning. Animals may constantly be seen to pause, deliberate, and resolve. It is a significant fact that the more the habits of any particular animal are studied by a naturalist, the more he attributes to reason and the less to unlearned instincts."*

This comment, made more than a 100 years ago, points to the idea outlined in the previous paragraph. Darwin had keen intuition concerning these aspects about which, in his time, there was not enough information but they were not solved to his complete satisfaction.

Primates show more propensity to remove themselves from the stimuli of their immediate environment and dedicate the majority of their cerebral activity to the problem they are considering. This is the reason why primates are considered the most intelligent land animals other than man.

Human beings demonstrate an extraordinary ability to concentrate on a singular problem to such a degree that they may ignore all the other happenings that are taking place around them and, in fact, block receipt from their other senses while they are concentrating. This profound concentration may endure for long periods of time. Of course, this is an extreme type of concentration; an intermediate type of concentration is more common in which the senses are not inhibited in their receipt of information while the individual concentrates. This intermediate concentration would be, for the most part, what we find in our daily activities in our working place. A secretary may be absorbed in typing a letter, but this does not prevent her from hearing the telephone when it rings (page 48 and 49).10

It is my opinion that this extraordinary power of concentration that is characteristic of man probably originated when man began to live in a group. As long as an animal is solitary, it is obligated to maintain its senses in a super-alert state, to be able to perceive any indication of danger in its immediate environment. When animals live in groups, as in the case of many plains animals such as the herbivores, this constant vigilance may relax somewhat, as there are many animals around the edge of the herd that are capable of communicating alarm to the rest of the group. As long as some are on the lookout, the rest may relax their senses somewhat and become involved with other activities. This is how, most likely, man found himself with more time to reflect on problems and solutions as he began to live in families, groups, villages, and so on. His intelligence developed as his brain became more and more efficient (page 49).11

If we observe an infant, we will notice that, at first, he doesn't have the power of concentration, and he is immediately distracted by any circumstance. As he begins to grow, little by little he is able to pay more attention to individual objects. As he gets older and his brain begins to function more efficiently, his power of concentration will also grow. With his power of concentration, he will begin to learn, store information, and make use of this information. Gradually, his intelligence develops. Power of concentration is, in fact, the basis of culture, as people acquire culture by imitating and listening to the adults around them and through formal education in schools (page 49)12 in most advanced societies.

Darwin says (page 460): *"It is generally admitted that the higher animals possess memory, attention, association, and even some imagination and reason."* Here, Darwin has changed his mind and admits that imagination is not a prerogative of man, as he said previously. He continues saying: *"If these powers, which differ much in different animals, are capable of improvement, there seems no great improbability in more complex faculties, such as the higher forms of abstraction, and self-consciousness, etc., having been urged against the views here maintained that it is impossible to say at what point in the ascending scale animals become capable of abstraction, etc.; but who can say at what age this occurs in our own children? We see at least that such powers are developed in children by imperceptible degrees."*

I agree with Darwin, in that animals learn by their own experience and also by observing the experiences of others. Many animals learn how to avoid being trapped or poisoned and not by their own experience; otherwise, they could not learn once trapped or poisoned. They acquire this experience by observing the outcome of such situations on other members of their species. That means that they observe, reflect, and, later, act accordingly to the results of such reflection stored in their mental computer, the brain. That means that they possess memory.

The anthropoids have a limited power of concentration, in that they are very easily distracted from the problem they may be considering. This has been well demonstrated in those studies where chimpanzees or gorillas have lived with humans and tried to adapt to their lifestyle. Another example of animals learning with little power of concentration is illustrated by animals used in circus and in movies. Their power of concentration is limited and therefore their ability to store experiences and information is also limited (page 49).13

I have emphasized the importance of man's power of concentration because without it man's cerebralization would be useless, as he would not be able actually to use the information he stores. The power of concentration could be considered the catalyst of culture as only in its presence are we able to make use of the resources stored in the human brain (page 49).14

Notes

1. *Man the Paradoxical Primate*, Jordi Fuentes, 1985.
2. Ibid.
3. Ibid.
4. Ibid.
5. Ibid.
6. *Contributions to the Theory of Natural Selection*, 1870, p. 212.
7. *Man, the Paradoxical Primate*, Jordi Fuentes, 1985.
8. Quoted by Sir C. Lyell, *Antiquity of Man*, p. 497.
9. *Man, the Paradoxical Primate*, Jordi Fuentes, 1985.
10. Ibid.
11. Ibid.
12. Ibid.
13. Ibid.
14. Ibid.

XIX

Division of Labor Determined by Sex

The nesting place or family nucleus is what obligated the division of labor determined by sex in hominids and early man (page 50).1 How can I be so sure that division of labor determined by sex existed already in hominids? Let us reflect a little about this point. First, we know that division of labor determined by sex exists in many animals, including mammals and birds. In many species of mammals, only the female has to care for the babies and teach them survival skills. There are species of mammals where females, whether they are the mothers or not of the offspring born in their group, clan, or family, take care of the young and teach them how to behave in adult life. Birds are not an exception, as there are many bird species in which only the female takes care of the young. The role of the male is only to fertilize the female, and after that is accomplished, they have no other intervention in the raising and protecting of the newborn.

In the case of humans and their predecessors, the hominids, the female had to nurse her young at regular intervals, look after his needs, and allow him to sleep during long periods of time. These obligations demanded that she not stray far from the nesting place and that she concentrate on those tasks she could complete within close proximity to her baby. She most likely would have dedicated herself to the collection of vegetables and fruits, and, if she lived close to the seashore, to the collection of shellfish. This relatively sedentary life led by the female could have meant that she very likely was the discoverer of agriculture, as she had a permanent or semi-permanent area available to her and the time to experiment with it (page 50).2 This situation infers different duties between both sexes. But, there is more.

The male did not have the same obligation to remain close to the nesting place as did the female, so he would be freer to dedicate himself to hunting animals, which would require that he be absent from the nesting place for certain periods of time, sometimes for days. After being successful on his hunt, he would return to the group so that they all might benefit. Only the male member of the group or family nucleus would have the autonomy to enable him to pursue this type of hunting (page 50).3

Undertakings such as looking for better areas in which to live and for better food availability, searching for water, and even engaging in conflict with rivals would be activities dominated by the male of the species. This characteristic has perpetuated itself until our time (page 50).4

In the case of the female, being limited to those occupation that she could pursue without neglecting her continuous maternal obligations gave her the opportunity to experiment with materials and to have a place to accumulate items that had a useful function for her, such as tools and utensils. She had the opportunity to try different materials for different functions, and adapt them to suit her needs. Most likely it was the female of the species who discovered the use of animals pelts as protection against the cold. She would have been the inventor of weaving, clothing, ceramics, and cooking variations or recipes. The male of the species, on the other hand, would most likely not have had this free time to experiment with new tools, foods, etc., and his need to be on the move while hunting would impede the carrying of tools other than the essential ones for the hunt (page 50).5 Evidently, the use of hunting tools forced the male to experiment in order to improve them and their efficacy.

It is very possible, following this same line of reasoning, that the female would have begun raising livestock. Remaining in one location would allow her to take care of orphaned animals and allow them to reproduce, and she would initiate the domestication of various species of animals (page 50).6

If we continue to analyze the advantages that remaining close to the nesting place offered the female, it would be very likely that it would be she who would have discovered the use of herbs, fruits, and bark for medicinal purposes. The male injured in a hunt would need to stay in the nesting place or home under the care of the female until he could once again resume his primary occupation. Women most likely were those who would have instructed or educated the children, at least until the male children were old enough to also participate in the hunt (page 50), 7 receiving, at this point, instruction from the father and male companions.

As we can see, the implications of the nesting place have been many and far-reaching. **These implications are the foundation of our present culture and our modern society.**

The division of labor according to sex is all too well known by present-day man. Up until just a few decades ago, the division of labor was very marked and distinct. At the beginning of this century in our Western culture, the thought that women could be involved in occupations such as mining, engineering, ship captaining, or construction, or could have the ability to wield a machine gun would have been dismissed as ludicrous. In other cultures, those activities considered a male prerogative are completely forbidden for females. The distinction or division of labor according to sex is very radical and actively enforced.

Presently, there is a tendency in Western culture for females to become involved in those areas traditionally dominated by the males of the species. On the other hand, males are presently demonstrating a tendency to participate in areas traditionally dominated by the females of the species. This tendency to become involved in areas traditionally dominated by the opposite sex is more pronounced in females undoubtedly influenced by economic factors, in whereby women now feel the need to play a role equal in economic reward to the men of our Western society (page 50).8

Notes

1. *Man the Paradoxical Primate*, Jordi Fuentes, 1985.
2. Ibid.
3. Ibid.
4. Ibid.
5. Ibid.
6. Ibid.
7. Ibid.
8. Ibid.

XX

Major Differences Between Man and Other Primates

With regard to the possible differences between man and the other primates, Darwin states in his chapter on *"Mental Powers"* (page 446): *"My object in this chapter is to shew that there are no fundamental differences between man and the higher mammals in their mental faculties."*

As far as Darwin had not defined what he understood by "higher mammals," it is difficult to approve of or disapprove of this statement. Today we consider whales and dolphins (Cetacea) as some of the most intelligent animals on Earth. I doubt that Darwin was referring to them when he used the term "higher mammals." It is more probable that he was considering as higher mammals only the anthropoids. This is unfortunately an obscure situation. If this is the case, I can agree only partially with his opinion. I personally accept, and most of the scientists do, that anthropoids have a high and advanced development in their mental faculties, and in many cases, they are at the same level as the human mental faculties. They can reason, reflect, and arrive at solutions that require a mental process, which is more complicated than the simple one of cause and effect. But (in science, there is always a "but") do the anthropoids possess the faculty of abstract thinking? We don't know. Should they have the faculty to produce abstract thoughts and ideas, I would agree with Darwin's statement, but if we cannot prove the existence of this modality of thinking, I cannot agree with the statement that *"there are no fundamental differences between man and the higher mammals in their mental faculties."*

Later, Darwin states (page 446): *"As man possesses the same senses as the lower animals, his fundamental intuitions must be the same. Man has also some few instinct in common, as the self-preservation, sexual love, the love of the mother for the new-born offspring, the desire possessed by the latter to suck, and so forth."*

Here again we face the difficulty in understanding what Darwin meant by "lower animals." He says that these lower animals must have the same fundamental intuitions as man, because we possess the same senses. According to the definition given in *Webster's Dictionary* "intuition: is: *"he immediate knowing or learning of something without the conscious use of reasoning."* Talking about the senses, Darwin says that "man possesses the same senses as the lower animals. . . ." Did he consider fish to be lower animals? If so, sharks have special organs that detect electrical currents of very low intensity. We humans do not possess this sense. Fish, in general, can detect changes of pressure in the water through specials organs along their bodies. We do not possess this sense. Some snakes have special organs to detect heat, and, with it, they can locate their prey in complete darkness. We do not possess this sense. Perhaps he included insects as lower animals. Bees can detect the ultraviolet color, but we cannot. Insects can detect changes in the atmospheric pressure, but we cannot. Do we possess the same senses as Darwin asserts? Definitely not.

We know of some animals that could hardly be considered "lower animals" that have senses we don't possess; for example, the whales and dolphins (*Cetacea*) that have the sense of sonar (sound navigation ranging) that allows them to swim and locate food in complete darkness. Bats are another example of animals with a sense that we do not possess, eco-location, similar in operation to the sonar of the Cetacea. Bats emit ultra sounds, which our sense of hearing cannot

a b c

The three largest anthropoids
a) chimpanzee (Pan troglodytes), b) gorilla (Gorilla gorilla), c) orangutan (Pango pygmaeus)

detect, that allow them not only to fly in complete darkness, but to locate their food without making use of vision.

Where do we go with all these observations? The immediate consequence is that humans **do not possess** the same senses as the "lower animals" as Darwin took for granted. We possess fewer senses than the "lower animals."

These examples are not used to criticize Darwin. I am not criticizing him, we cannot do that, because today we know of the existence of these mentioned senses found in different species. These senses were not known in Darwin's time, so we cannot consider this lack of information as his oversight. It was not his fault. It is the same situation that occurred with the laws of heredity; Darwin did not know of their existence. He did not know the existence of these specialized senses. I am pretty sure that even today we don't know of all the existing senses in the animal and vegetable world. It is very possible that nature has many more surprises for us. We are very ignorant in regard to the biosphere around us, the Earth, the "Pachamama" or Mother Earth of the Quechua Indians.

As everything is changing, so is our information. It is very possible that in a few years, all that we know today will be obsolete and substituted by new findings and discoveries. Our conception of the universe, our world, and ourselves will be very different from the view we have today.

What Separates Man from the Other Primates?

Man has many characteristics that separate him from other apes and primates, but only the most obvious will be mentioned (page 11).1

1. Human babies are born in a helpless state, and this fact obliges man too adopt a bipedal position; the bipedal position obliges erect posture.

2. Man is the only nidicolous primate.

3. Man is the only primate that exhibits the characteristic of neoteny. Neoteny is the retention of juvenile or immature characteristics during adulthood. Adulthood is the completion of physical growth and the ability to reproduce oneself. Human beings, because of this state of neoteny, are able to reproduce before acquiring adult physical characteristics. Human beings normally develop until the mid twenties but are able to reproduce themselves in some cases as early as ten years old for females and twelve years for males.

4. Man is hairless. This is another fact that obliges man to use his arms to hold his offspring as the offspring cannot cling to the mother in the manner of apes and monkeys.

5. Man is the most cerebral of the anthropoids.

6. Man is the only carnivorous primate that has not developed carnivorous teeth.

7. Man possesses a highly developed spoken language, and he can sing and whistle. Orangutans only produce certain rhythmical sounds during courtship.

8. Man possesses an opposable thumb and very sensitive fingertips.

9. Man's foot does not have characteristics that would suggest the human foot was, at one time, prehensile and used in arboreal life, as is the case with other anthropoids. Rather, his foot possesses characteristics that are adapted to terrestrial life.

<u>Man's foot has always been a foot, never a hand.</u>

10. Man's power of concentration is unique.

11. Man is the only primate that can laugh and cry.

12. Man can mate in both frontal (*hominum*) and posterior (*canum*) positions. Bipedalism places the female sexual area midway between her front and her back. Chimpanzees are the only anthropoids that sometimes mate in the *hominum* position.

13. Man is the only primate whose female mammary glands are not significantly reduced in volume once the nursing period is over.

14. Man is the only animal with a developed obsession to accumulate things that are not of

immediate use for survival, such as richness and honors. Man has developed the condition of greed.

15. Man is the only primate able to change his environment even to the point of destroying his own species and all life on Earth.

Note

1. *Man, the Paradoxical Primate*, Jordi Fuentes, 1985.

XXI

Comparison of Primate and Human Proteins

There have been phylogenetic studies on primates of the reactivity of eleven monoclonal antibodies to human T cells and rosette formation, in the reaction of primate T cells with sheep erythrocytes, demonstrating that the sheep erythrocyte receptor and the determinants of T cell subset antigens are highly conserved during primate evolution, while other T cell antigens are less well conserved. Identical reactivity was shown by human, gorilla, and chimpanzee T cells with the monoclonal antibodies to human T cells. Additional evidence that man and African apes shared a relatively recent common ancestor is provided by the expression of human T cell antigen determinants by primate T cells, suggesting a constant rate of evolution for this group of molecules. 1 These experiments, carried out by Barton F. Haynes, Barry L. Dowell, Lucinda L. Hensley, Ira Gore, and Richard S. Metzgar of the Department of Medicine, Division of Rheumatic and Genetic Diseases and the Department of Microbiology and Immunology, Duke University Medical Center, Durham, North Carolina, suggest, through fossil studies and biochemical information, that humans, chimpanzees, and gorillas have a common ancestor approximately five million years ago. By contrast, the common ancestor of man, African apes, and the orangutan goes back about eight million years ago. The common ancestor of man, African apes, the orangutan, and the gibbon is about ten million years old. It is also estimated that the Old-World monkeys and the apes separated approximately thirty million years ago (page 55).2

The ancestors of the New World monkeys would have to be separated prior to 250 million years ago, as at that time, Africa and South America began to separate already. That corresponds to the separation of platyrrhines from the catarrhines. One hundred million years ago, both continents had practically the same separation as today. The Atlantic ocean separates the two continents at a rate of 1.5 inch per year, approximately (page 55).3

I have always found it very difficult to believe that animals could have populated lands separated by great bodies of water by drifting on tree trunks until running into land. This could only be possible with the occurrence of various highly improbable circumstances. First, in order for a species to be able to populate a new area, both a male and a female of the same species and of the proper age and in the proper condition to reproduce themselves must arrive at the same area within a contemporary time period (page 55 and 56).4

Possibly, a pregnant female could drift to a new area. In this case, however, it would seem more likely that during a voyage of such magnitude and length, as is crossing the Atlantic Ocean, the pregnant female would give birth during the voyage. If, in fact, the female did give birth during this voyage, what possibilities are there that the offspring and the mother could survive without food, water, or a way to protect themselves against the elements (page 56)5 ?

The similarities between animal and plant life between South America and Africa are so great that it would oblige one to believe that these voyages across the Atlantic Ocean would have to have been made by plants and animals alike, with the plants establishing themselves on a log well enough to complete a transoceanic voyage. How many thousands and thousands of drifting logs would be necessary, all of them with enough food and water to support the life they carried, for even one to arrive at a new continent? It would have to have been a continuous fleet of logs moving for hundreds of years, fighting currents running parallel to continents. As you can see, I consider this to be a basically absurd explanation of the origin of South America's animal and plant life (page 56).6

The probability of terrestrial animal migrations over large bodies of water would seem to be valid only when the distances are short or when the animals could actually swim to their destination, as do moose and polar bears. Only the bird, making use of flight, is able to cross large bodies of water to new land areas successfully (page 56).7 One example is the colonization of America by the cattle egret of Africa (*Bubulcus ibis*), whose presence on the American continent began in the last century in British Guiana, and a few were breeding in Florida in 1953. Today this bird is a very established resident in North America, reaching as far north as Canada.

The information gathered during the studies mentioned before seems to confirm the original idea set forth in this study, that man occupies the summit of a central evolutionary stem that continued in a straight line, characterized by newborns being born in a helpless state, forcing bipedal locomotion. Lateral branches of this main trunk would have been produced over a long period of time. The first branch was the separation of the platyrrhine monkeys (belonging to the infraorder *Platyrrhini*). The second branch would have been the separation of the catarrhine monkeys (belonging to the infraorder *Catarrhini*). The third would have been the Driopithecines, which was an unsuccessful branch that became extinct between fourteen and ten million years ago. Next appeared the division of the gibbon branch. They were anthropomorphic. This took place approximately ten million years ago. About seven million years ago the orangutans had separated, finding their evolutionary niche (page 56).9

It is generally accepted that approximately five million years ago the chimpanzees and gorillas separated from the ancestor of the humanoids. These two anthropoids have remained together until present times; even though they have been considered separated into two different genuses, it is now believed that the genus *Gorilla* and the genus *Pan*, both may belong to the same genus. I don't agree with that; but this is a job for Taxonomists. From this date and perhaps even before, man has no longer shared ancestors with the anthropoids (page 56).9

The division of the anthropoids from the central line of evolution would be a consequence of their offspring being born with coordinated, conscious reflexes, with prehensile feet and the ability to sustain their own weight from birth. This allowed them to occupy evolutionary niches heretofore vacant (page 56).10

The central evolutionary line continued its development through the australopithecines in different varieties, some of which became extinct, until the appearance of the genus *Homo* which, from *Homo habilis*, through *Homo erectus* and after *Homo sapiens*, arrived at the present state. We consider ourselves a subspecies, *Homo sapiens sapiens* (page 57).11

This proposed sequence seems to be logical, given our present knowledge. It is possible that in the future, with the discovery of more fossils of our ancestors, our accumulated knowledge will augment and we will be able to pinpoint more evolutionary stages and find intermediate species and our true ancestors. Perhaps, if luck accompanies our findings, we will be able to determine the lateral species that were related to us but which disappeared without taking part in the formation of our species (page 57).12

Discrepancies

Some doubt is held regarding the comparisons between primate and human proteins. The main argument is that while these comparisons may be a method to find our common ancestors in various species, the present technical state of the research is not well enough developed to be completely reliable. It is possible that these doubts may be well-founded, but until we have other methods of comparison, this is the one we will have to use (page 56).13

The First Life on Earth

Until recently, it was thought that the necessary chemical reactions to produce the first life would have taken place in the sea. However, there is now another hypothesis that perhaps these chemical reactions took place in clay on the land, and not in the sea.

The main reason why this theory has been presented is that it is believed that the chemical reactions needed to create life include the loss of water molecules, and it is very improbable that this loss could take place in an aqueous environment. There are also other important considerations. (page 56).14

Ramapithecus

Due to the latest fossil discoveries found in the South of China, the problem of the placement of Ramapithecus in the human line of evolution has arisen. Until these discoveries, it was believed that Ramapithecus was in the human evolutionary line; but with the latest findings, it is now thought that Ramapithecus is, in fact, an ancestor to the orangutan, and is no longer thought to be a direct ancestor of man (page 57)15.

Notes

1. B. F. Haynes et al., "Human T Cell Antigen Expression by Primate T Cells", *Science*, January 15, 1982, pp. 298 and 299.
2. *Man the Paradoxical Primate*, Jordi Fuentes, 1985.
3. Ibid.
4. Ibid.
5. Ibid.
6. Ibid.
7. Ibid.
8. Ibid.
9. Ibid.
10. Ibid.
11. Ibid.
12. Ibid.
13. Ibid.
14. A.G. Cairns-Smith, "The First Organisms," *Scientific American*, June 1985, pp. 90 to 100; *Man, the Paradoixical Primate*, Jordi Fuentes, 1985, (see p.56).
15. *Man, the Paradoixical Primate*, Jordi Fuentes, 1985.

XXII

Is Man Really a Primate?

Fortunately, science is not dogmatic, unlike some other expressions of human thought. Science is self-critical, and accepts analysis and correction.

If we summarize human physical characteristics as well as human behavior, we find ourselves facing the paradox that man is a primate unlike all other primates. Primates are not classified according to a special physical characteristic, since primates are not anatomically specialized as are other animals. Rather, primates are classified by a series of characteristics that may be present to some degree or another in other mammals. In other words, the order *Primate* includes all of those mammals that cannot be classified within the already established orders. Humorously, the order *Primate* is the trash can for those creatures that cannot be included in any other order.

Within the order *Primate* the human being is placed within the suborder *Anthropoide*, the superfamily *Hominoidea*, which is shared with the anthropoids, and, going one step further down, he is placed in the family *Hominidae*, in which only he and his closest ancestors are grouped.

If we make a detailed mental analysis of the characteristics representative of this family, we will see that the differences found between human beings and other primates are many and outstanding, and one would begin to wonder if man has been properly classified.

Darwin was not comfortable with his own conclusion about the supposed relation of man with the other primates. In his chapter on *"The Mental Powers,"* he said, *Man bears in his bodily structure clear traces of his descent from some lower form; but it may be urged that, as man differs so greatly in his mental power from all other animals, there must be some error in this conclusion."* This demonstrates that he insinuated or held suspicions that perhaps man did not descend from the apes, letting the door open to another unforeseen possibility.

In the eighteenth century, the famous botanist Carolus Linnaeus (Latin for Carl Von Linné, 1707-1778) developed the binomial system that is still in use today, and he established taxonomy as a discipline.

Up until present-day time, it has not been questioned that the universally accepted classification of human beings could be wrong. The constant use of this classification through the years has converted this taxonomic classification into a near dogma that is accepted without discussion. But . . . . is it right?

I am not suggesting here that this classification is absolutely erroneous, and I propose no new solutions. It is possible that it is correct, or, for the moment, the most correct, given the information we have at hand. What I am proposing, however, is that, at this time, enough material exists to warrant a profound study into the exactness of man's present classification, to try to determine whether man has enough peculiar characteristics to establish his own order.

Recognizable Characteristics Used to Cassify an Animal under the Order Primate

Foot posture plantigrade.

Soles of feet naked, with enlarged pads.

Nail always present on hallux, usually also on other digits.

Pollex and/or hallux opposable, used for grasping.

Braincase relatively large, housing well-developed cerebral hemispheres.

Radius and ulna, tibia, and fibula separate.

Clavicle well developed.

Orbit large and separated from temporal fossa by a postorbital bar or plate.

Molars trituberculate or quadrituberculate.

XXIII

Conclusion

Now that we have made a brief analysis of the characteristics that separate man from anthropoids, we can recount some of the situations that clarify the conclusion to the original idea concerning this work (page 59).1 The foundation of all the ideas and hypotheses set forth in the preceding chapters is the helpless state in which human beings are borne and, by deduction the helpless state in which our ancestors were born, forging the basis for the existence of our species and eventually our present-day civilization. Our lineage, as it is known today, most likely originated many millions of years ago at the time of the first nidicolous primates. Some of the features and characteristics of these nidicolous primates that were developed in the previous chapters are given here (page 59).2

The helpless state of the newborn obligates the parents to develop a terrestrial and bipedal way of life (page 59).3

Bipedalism is, in my opinion, a technological innovation of reproduction, not locomotion.

The appearance of neoteny allowed the species to survive (page 59).4

Man is the only nidicolous primate, and the need to have a nesting place where he may leave and care for his offspring, and where the family could rest, gave birth to group living, which eventually led to the town and the city (page 59).5 Group living imposes certain obligations on each sex, and therefore division of labor according to sex is its consequence (page 59).6

The cerebralization of the *hominid* gave rise to the evolution of *Homo* and the development of spoken language and abstract ideas. Language most likely appeared at approximately the same time when group life and bipedalism were evolving. Language facilitated group living, which eventually developed into the village and the city. Language also facilitated division of labor. Technology appeared sometime after the development of language. With the development of group living, the necessity to have order to it arose, and this order was found in the hierarchical structure of the individuals composing the group (page 59).7

This human evolutionary path arises from a direct line of the first mammals with simian characteristics, which would have appeared more than 100 million years ago. This original primate or simian would be terrestrial and would posse tendency to give birth to a helpless offspring, therefore becoming nidicolous. Neoteny would appear together with nidicolous family organization as a way of raising the young. The group or species that possessed these characteristics would be the center line in the evolution of man. Lateral lines developing from the central evolutionary line would be first the catarrhines, then the Driopithecines, then the gibbons and orangutans, and later the gorillas and chimpanzee (page 59).8

The explanation as to why it is more logical to assume that the human ancestor was the central line of evolution and not a lateral line, is that for the purpose of natural selection, it does not make sense that a species would developed with able newborns, alert and capable of a certain degree of autonomy, to then go back to develop newborns less able to protect and take care of themselves. If this were the case, evolution would be working toward a species less able than the original, which does not make sense and which would probably spell out the extinction of the species. It is much more likely that human ancestors carried on the central line of evolution slowly and steadily until arriving at the genus *Homo*. This evolutionary path would be distinguished by a specialization in the area of cerebral development, with all of the accompanying implications and advantages of this development (page 59).9

The simians first, and later the anthropoids, were those lateral branches that developed newborns better able to take care of themselves, with better coordination and reflexes, and occupied available ecological niches. With these ecological niches occupied, there there would be no need for greater cerebralization for the survival of the species, they would continue to differentiate themselves from the central evolutionary line (page 59 and 60).10

Indeed Darwin's ideas concerning evolution are vital to understanding living creatures today, and to explaining many situations that were disjointed and lacking explanation in Darwin's time. Perhaps his enthusiasm led him astray when he considered that man was closely related to the monkeys, as he frequently associates both throughout his work. He was very careful not to take a strong position, so he said (page 444): *"In regard to bodily size or strength, we do not know whether man is descended from some small species."*

Perhaps Darwin, as well as his followers, fell into erroneous thinking, considering that man was the ultimate creation of nature, based only on his civilization; and as a logical consequence, he must have originated from existing species considered to be older. Misinterpretation of the situation existed in Darwin's time, and exists in the present. Human civilization has been expanding for approximately the last 10,000 years, until reaching the present state of sophisticated technical evolution including expansion of the culture and communications around the world. Spatial exploration and the global union of people via the Internet are in reality the most recent accomplishments on this planet. This though produces the image that, as our civilization and technical development are the most recent achievements, the species that produced it also has to be the last arrival to the biological concert. This is not reality. There are many species older than man on Earth, that have not progressed as we did. Crocodiles and turtles are millions of years older than man or Earth. What does that mean? It does not mean that, because we have accomplished late technological developments our species is a recent arrival. Our species and the ancestors from which we have evolved were on this planet for several millions of years without any spectacular technical progress. The image that man is the last arrival is produced when we consider human civilization as the most recent event on Earth; this has produced the idea that our species is also a recent arrival. Thinking about this situation, it is logical to arrive at the conclusion that we are a recent arrival, then our species has to be an offspring of some other existing species; the closely related ones are the anthropoids. This is what spurred the idea that we have evolved from them. This is not the case. Our civilization is a recent event, no doubt about it, but our species and our ancestors were on Earth millions of years before human civilization sprouted.

Our species is not a recent arrival, only our civilization.

Darwin's revolutionary ideas have caused the same amount of contentment among scientists as they have indignation to clergy. The same debate that ensued in Darwin's time continues today between the "creationists" who, despite existing evidence based on modern medicine, genetic science, agriculture, and livestock raising, cannot accept the idea that we are an animal subject to the same natural laws as other species of animals (human arrogance at work) and the "evolutionists," supporting (totally or in part) Darwin's theory.

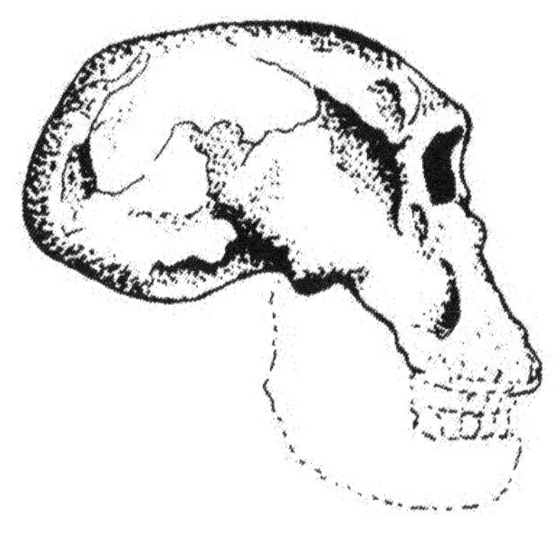

Homo habilis

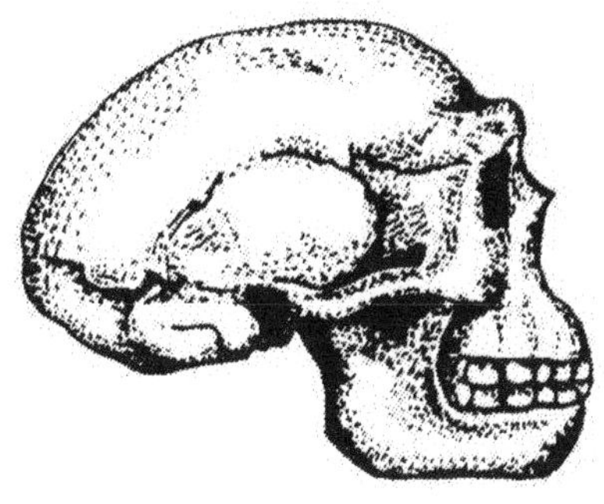

Homo sapiens

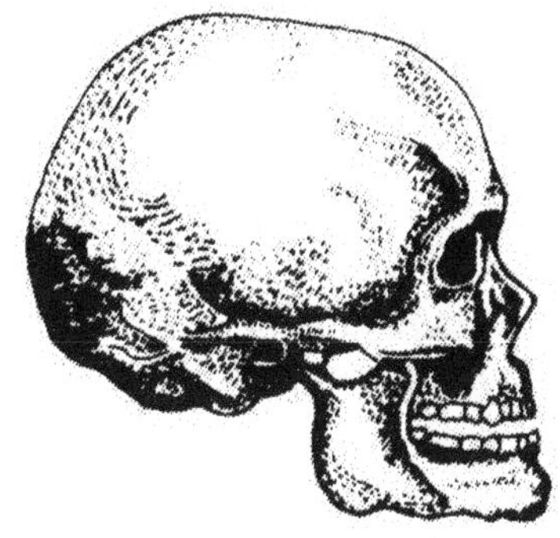

Homo erectus

This resistance to accept reality originates in the image that one conjures of our ape-like ancestors and our reluctance to claim any relation to "inferior" species such as the monkeys, animals considered comical and hardly worthy of dignity and respect. So it is more appealing to accept our origins as an act of deliberate creation. That way, we avoid any possible connection with these species that we despise. Human arrogance has found a simple and straightforward solution to an otherwise complicated situation.

I believe that man is related to the primates and, more remotely with all mammals, but that he has carried on a central line of evolution a very old line of evolution, that eventually led to a being who could modify raw material to suit his needs. Other related species differentiated themselves from this central line occupied by our ancestors (page 60).11

One fundamental characteristic that distinguishes and separates man from other animal is his ability to transform raw materials into specific tools for specific purposes. With the continuous creation of new tools to modify his surroundings, man advances in the understanding and dominion of his environment. These achievements man owes to his cerebral development, which permits him to store a large amount of information in his memory and, at the same time, allows him to originate new ideas and concepts that he communicates through a complex spoken language ancestors (page 60).12

Had it not been for the ability to transform raw materials into other products, man would still be in the Stone Age, and he would not be so distinct from his anthropoid brothers, as is possible that he has been for thousands and even millions of years. The ability to transform raw materials gave birth to the moment when man learned to use fire. The first related transformations were most likely in the area of the preparation of food, offering variety and different combinations of food not previously tried. Using sticks to roast meat probably let to the discovery that wood can be hardened by the action of fire, and with this discovery came the possibility to make more durable spears and arrows. Fire permitted the use of hard and poisonous foods, otherwise inedible. In experimenting with fire, later man would learn how to make medicines and dyes by means of cooking vegetables and minerals. The next step was the production of metals by means of the action of fire. Fire also afforded him protection against other animals and the cold weather; it later provided the ability to use new areas for planting by burning a field prior to planting. The ability to burn fields and to plant would encourage the expansion of agriculture (page60),13 and with greater production of agricultural products came an improvement in the possibilities of survival and enlargement of the human group.

Intelligence, as defined in anthropology, is the faculty to find a solution to a problem. He who is capable of finding the solution faster is considered more intelligent than he who takes longer. When we refer to *"intelligence"* as the ability to solve a new problem, most of us are inclined to think that we are in possession of all the intelligent resources available, or we think of ourselves as having the accumulated intelligence developed through thousands of years. We, in fact, do not possess this type of intelligence, but, rather an ability to store knowledge. The accumulation of knowledge passed from one generation to the next is what we know as *"culture"* not intelligence. Our intelligence gives us access to this culture and allows us to make use of its knowledge. "Culture" is defined in anthropology as the knowledge transmitted from parents to children; in a wider scope, the knowledge transmitted from one generation to the next.

The majority of us learn to use complicated machinery with a little instruction, but the fact that we can use the machinery does not mean that we understand how to make such a machine. If we push a button and cause the television to turn on, we are, in fact, doing the same type of thing that a chimpanzee is perfectly capable of doing. Even a seriously mentally retarded person can do it.

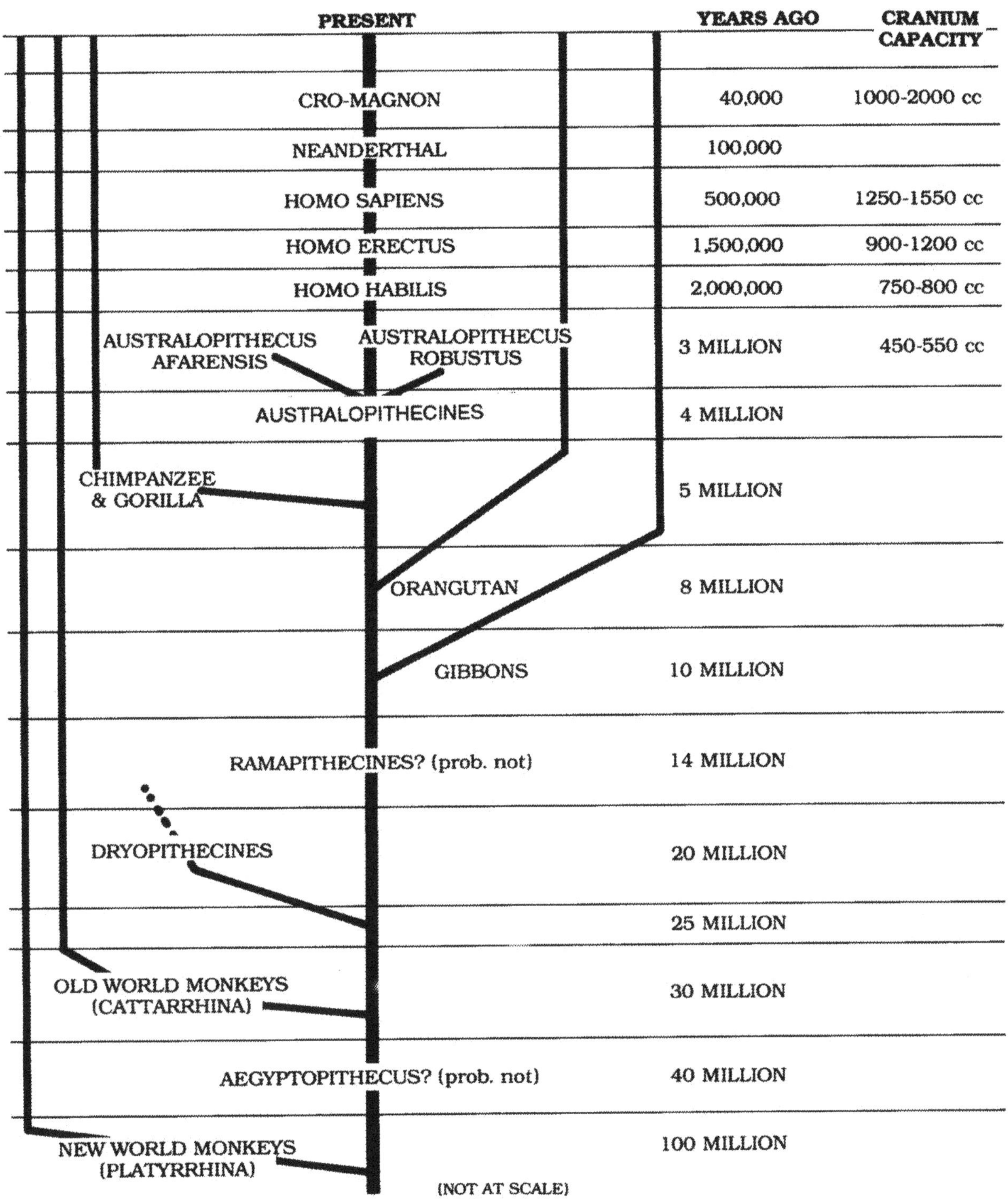

Diagram of a possible family tree for primates. The central trunk is a direct stem to the present Homo sapiens sapiens . The "tree branches" show the possible split off from the central trunk followed by the other primates. The first branch is that of a New World monkeys (Platyrrhini) that had to be separated from the main trunk more than a hundred million years ago. At this time the Atlantic Ocean was already formed, separating South America from Africa.

Intelligence is not shared equally. Some people are more intelligent than others, that is, they are more capable of finding solutions to new problems. But as mentioned before, this ability to solve new problems should not be confused with culture or experience. Intelligent persons could have very little culture; persons with much culture may have little intelligence. In the latter case, the accumulation of knowledge and experience can hide the fact that a person has little intelligence. The person who has learned from others can apply learned solutions to problems, and it looks like he is more intelligent than he actually is. He is using experience, not intelligence (page 61).14

Accumulation of culture is the result of the use of individual intelligence, but the totality of knowledge is found in the group, not the individual. Each individual brings his grain of sand (some, a rock) to the culture of the group. If we were to add up the cultures of all diverse groups, we would have the sum of human culture, the product of the activity of all the intelligence that has existed since our oldest ancestors began to accumulate knowledge and experience; at the same time we would be manifesting the ability to use this knowledge when needed (page 61).15

I have put forth with detail, and perhaps with some redundancy, the most outstanding features of the evolution of our species, giving explanations to situations not understood, answers to various questions, and solutions to some problems presented in anthropology. It is very possible that I have produced more new questions than answers to old ones, but this is the way that we advance in our search for explanations.

All matters treated in this work are the product of careful reasoning based upon a great accumulation of reliable, if indirect, information. This information has been the basis for arriving at satisfactory conclusions and explanations concerning points little understood. Explanations have been given here that have not been given before, not because of an oversight of the individuals involved with this material, but because of a lack of connection between two complementary sciences: biology and anthropology (page 61).16 Of course, it is possible that the conclusions presented in this work are not 100 percent correct, but I think they may be close, and are worthy of further investigation and study. In addition, as Dr. Isaac Asimov says,

"In science it is not so important to be correct (including there may be no way to determine what is correct), it is simply necessary to be sufficiently correct for the moment. (page 61 and 62).17

Notes

1. Man, the Paradoxical Primate.
2. Ibid.
3. Ibid.
4. Ibid.
5. Ibid.
6. Ibid.
7. Ibid.
8. Ibid.
9. Ibid.
10. Ibid.
11. Ibid.
12. Ibid.
13. Ibid.
14. Ibid.
15 Ibid.
16. Ibid.
17. Ibid.

Biographies

The biographies included in this section permit the reader to obtain rapid, concise information about those persons whose ideas, theories, and works have contributed to, or have exercised some influence on the development of the studies of evolution and present-day genetic science. It is possible that some readers will disagree with the selection, perhaps believing that some of those selected did not have enough merit to be included, or that others who were omitted were of greater importance and should have been mentioned. It is very possible, also, that some readers, mighy believe that I should have included those thinkers who are opposed to the idea of evolution. But, in a work on astronomy, should those persons who think that the Earth is flat, or that it does not move, be included ? These diverging points of view, differences, and criticisms are to be expected, as it is very difficult to meet everyone's criterion. For any oversight, I can only ask for your indulgence.

Aristotle
(384-322 B. C.)

Aristotle was Greek philosopher, educator, and scientist. He was born in Atagira, a little town in northern Greece. He was the son of Nichomacus, the personal physician of Amyntas II, king of Macedonia, who was grandfather of Alexander the Great. Aristotle was eighteen when he entered Plato's school in Athens, known as the Academy. He was the brightest student in the Academy. At about 343 o 342 B.C., he was the tutor of Alexander, the son of Philip II of Macedonia, who was known as Alexander the Great when he conquered Greece and overthrew the Persian Empire.

At about 334 B.C. Aristotle returned to Athens and founded a school called the Lyceum. He, his philosophy, and his students were known as the *peripatetic* (from the Greek word *Peripatos*, to walk) because he taught while walking.

Among his multiple interests, biology received special attention. Aristotle suggested the first classification of animals and plants. Only about thousand animals were known by then. He divided the animals in two groups: animals with red blood (animals with backbones) and animals without red blood (animals without backbones). This classification lasted for almost 2,000 years, until the eighteenth century, when Carolus Linnaeus established the binomial classification of animals and plants.

Crick, Francis H. C.
(1916 -   )

Francis H. C. Crick was born in Northampton, England, in 1916. He studied at London and Cambridge Universities. He is a biologist. In 1962 he shared the Nobel Prize in Physiology with biologist James D. Watson and biophysicist Maurice H. F. Wilkins.

Crick and Watson built a model of the molecular structure of *deoxyribonucleic acid* (DNA). DNA is the substance that transmits the genetic code from one generation to another. The shape of the DNA molecule is a double twisted spiral that is called the *Watson-Crick model*. This work has been fundamental to the understanding of the sequence of the genetic code.

Darwin, Charles Robert
(1809 – 1882)

Charles R. Darwin was born on February 12, 1809 in Shrewsbury, England. He was educated at the Universities of Edimburgh and Cambridge. After his graduation he sailed as a naturalist, to the scientific expedition aboard the H.M.S. Beagle, sailing in December 1831 and heading around the world.

During the five years that the expedition lasted, they sailed along the East and West Coast of South America, and explored the Galapagos Islands and other islands of the Pacific.

This exploratory trip was a field laboratory for Darwin, who compiled important research and collected fossils and specimens of plants and animals. His observations were the basis for his theory about evolution through natural selection, which he explained in his works. His book *On the origin of the Species by Means of Natural Selection, or the Preservation of Favoured Races in the Struggle for Life,* published in 1859, caused a revolution in natural sciences.

The storm of debate that this publication produced increased when he published his other book tittled *The Descent of Man* (1871). At this point, his works began to receive hard criticism because people were not prepared to accept his hypothesis that man descended from the same group of animals as the gorillas, chimpanzees, orangutans, and gibbons, known as the Anthropomorphs or, simply, apes.

Soon after his arrival back from the voyage around the world, Darwin settled in London (1836), and in 1839 he married his cousin Emma Wedgwood. This marriage produced five sons. In 1842, Darwin moved to Down. He passed away in 1882, and was buried in Westminster Abbey.

Huxley, Thomas Henry
(1825 – 1895)

Thomas H. Huxley was born near London, England, in 1825. He entered medical school and became a surgeon in the British navy. He spent four years in the Indian Ocean and East Indies. He was interested in zoology, and he wrote an interesting work about jellyfish. Due to this work, he became a famous zoologist. From 1854 to 1885, he taught natural history at the Royal School of Mines. He served as president of the Royal Society from 1881 to 1885.

He accepted the theory of Charles Darwin and, through his lectures and writings, he helped to advance scientific thought. He was a very prolific writer of scientific subjects. He produced an essay titled *"On a Piece of Chalk"* (1868). He also wrote an essay in which he demonstrated the gradual evolution of the horse foot. His writings include *"Evidence as to Man's Place in Nature"* (1863), *Critiques and Addresses"* (1873), and *A Manual of the Anatomy of Invertebrate Animals* (1877). He was also well known due to his introduction of the word *"Agnostic,"* one who believes that the existence of God or a spiritual world cannot be proved, and the word *"biogenesis."*

Lamarck, Chevalier De
(1744-1829)

Lamarck was born on August 1, 1744 at Bazentin, Picardy, France. His name was Jean Pierre Antoine de Monet. He studied briefly for the priesthood. He served as an army officer during the Seven Years' War. After the war, he studied medicine (1768), but at twenty-four he began to study under the noted botanist Bernard de Jussieu.

Through his studies in botany, Lamarck concluded that plants and animals change their forms to adapt to their environment, and that theses changes are passed along to their offspring. He became one of the first to propose a theory of biological evolution. His studies on plants and animals helped Charles Darwin to develop his theory of evolution.

Lamarck became the conservator of the Royal Herbarium. Little by little, he changed his interest from plants to animals, and in 1793 he was appointed as professor of zoology at the Museum of Natural History in Paris. He developed a system to classify invertebrate animals.

Lamarck was interested in the study of animal fossils, and was the founder of invertebrate paleontology. He had much interest in natural history, and was the first scientist to try to forecast the weather. He published an annual meteorological report from 1799 to 1810. Probably he was the person responsible for the names of the clouds: *cirrus, stratus, cumulus,* and *nimbus*. In his later years, he was completely blind, but he continued his work with the help of others.

Leakey, Louis S. B.
(1903 - 1972)

Louis S. B. Leakey was born in Kabete, Kenya, in 1903. He was the son of English missionaries. He discovered many fossils that represent various stages of human evolution. He worked with his wife Mary; together they discovered in Kenya, fragments of a jaw and teeth considered to be fourteen million years old. According to the Leakey's research, these fragments probably represented an early ancestor of the human being; they named this possible relative *Kenyapithecus*. The Leakeys also discovered many other remains that they believed to belong to a more developed species, which they called *Homo habilis*. This last relative of our species was contemporaneous of the *Zinjanthropus*, considered a type of primitive man from East Africa. Those remains were discovered by Mary in Oldubai Gorge in Northern Tanzania.

Leakey, Mary Nicol
(1913 -   )

Mary N. Leakey was born in London, England. She was the wife of Louis S. B. Leakey and participated in his anthropological research in Kenya. With her husband, Mary discovered many fossils among them, some fragments of a jaw and teeth, considered to be fourteen million years old. This creature received from the Leakeys the name of *Kenyapithecus*.

Mary Leakey, while working in the Oldubai Gorge in northern Tanzania, discovered, in 1959, a human-like skull considered to be 1,750,000 years old. She named this creature *Zinjanthropus*. Later, this fossil was renamed *Australopithecus boisei*. Mary Leakey also discovered the first footprints of a group of biped hominids. Together with her husband, she discovered the remains of a new creature considered to be a close relative of the human being, which received the name *Homo habilis*.

Linnaeus, Carolus
(1707-1778)

Linnaeus was born in 1707 in Råshult, near Kristianstad, Sweden, with the name Karl von Linné. He became known as Carolus Linnaeus because he wrote his books in Latin. His father was a parish curate and wanted his son to study for the ministry, but the boy was more interested in plants. Their friends urged his father to send Karl to medical school. During his stay in medical school, Karl supervised a small botanical garden and began to become interested in entomology. He began to write about the plants that he knew with very accurate descriptions. These notes were the basis for his later books.

The Royal Society of Science gave him a grant of $50; with that money, he spent five months in Lapland collecting plants. Later, he went to the Netherlands, where he received his medical degree. Alter his return to Stockholm, he practiced medicine. In 1742, he became a professor of botany at the University of Uppsala.

Linnaeus was the one who established the present method of classifying animals and plants, inventing the binomial nomenclature system of giving a double name to each plant or animal, consisting of the name of the genus (group) followed by that of the species (kind). He published the book *Species Plantarun* (1753), in which he established the classification of plants. The classification of animals was established in his book *Systema Naturae* (1758). With these two important works, the modern classification of plants and animals was organized, making it easy to identify a plant or animal, regardless of the local or popular names given to them in different countries or locations.

Malthus, Thomas Robert (1766-1834)

Malthus was born on February 17, 1766 in Surrey, England. He wanted to be a clergyman and graduated from Cambridge University. He took over a parish in Surrey (1796). He became a professor of history and political economy at the College of the East India Company (1805) and held this position until his death.

Malthus is best known for his *Essay on the Principle of Population*, published in 1798. The idea exposed by Malthus in this book is that population tends to increase more rapidly than food supplies. As a consequence, he saw that the only solution to this situation is that wars and disease would have to eliminate the extra population, unless people decided to limit the number of their children.

Darwin found, in the ideas of Malthus, that a relationship existed between progress and the survival of the fittest. This was the basic idea in the theory of evolution established by Darwin.

Some scientists say that Malthus' prediction failed in the 1800s, as the industrial revolution provided better systems to increase the production of food supplies. However, the situation at the end of the twentieth century is heading towards the result predicted by Malthus, and there is renewed interest in his ideas. Some countries, like China, imposed drastic reductions in the growth of the population, and in many underdeveloped countries famine is controlling rapid population increase. Many conservationists warned that food production could not keep pace with population growth indefinitely. The solution proposed by *neo-Malthusians* is birth control, an idea that Malthus rejected.

Mendel, Gregor Johann
(1822-1884)

Gregor J. Mendel was born on July 22, 1822 in Heinzendorf, Austria. His father was a farmer, and Mendel became interested in plants on his father's farm. He entered the Augustinian monastery in Brünn in 1843. Most of his life was spent in the monastery, except for short periods while he was a student at the University of Vienna, and during his teachings of natural history at nearby schools.

Mendel conducted experiments for many years in the garden of the monastery; observing the contrasting characteristics of the different members of the same species of plants, he began to grow successive generations and took accurate notes of the results. From these experiments, he discovered the principles of heredity.

One of the plants used by Mendel for his experiments was the sweet pea; observing the characteristics of each successive generation, he proved that there was a definite pattern in the appearance of the different traits with regard to color, texture, and so forth. He found that it was possible to raise plants that showed only one characteristic, and that gave him the idea that there are two different kinds of characteristics, dominant and recessive. He could select a line that showed only dominant characteristics and another that showed only recessive characteristics. Through keeping meticulous records of the different generations, he discovered the principles of heredity. That was the beginning of present day genetic science and its application, genetic engineering.

Mendel published his experiments in 1866 in a paper printed by the Natural History Society of Brünn. Unfortunately the importance of his work was not discovered until 1900.

Watson, James Dewey
(1928 -)

James D. Watson was born in Chicago, Illinois. He studied at the University of Chicago and at Indiana University. In 1961, he became a professor of biology at Harvard University.

He shared the Nobel Prize (1962) with biologist Francis H. C. Crick and biophysicist Maurice H. F. Wilkins. Watson and Crick developed the *Watson-Crick model* of the structure of the deoxyribonucleic acid (DNA) molecule. DNA is the substance that carries genetic information from one generation to the next.

The model of the DNA molecule has the shape of a double twisted spiral, united by rungs of DNA, having the aspect of a ladder. This discovery was the key to understanding the sequences in the genetic code.

Glossary

This glossary includes only those definitions whose meanings are related to the subject of this work.

Glossary

Aminoacid	A group of nitrogenous organic compounds that are the units for the structure of proteins. They are essential to metabolism.
Ape	Received this denomination and of the anthropomorph primates: orangutan, gorilla, chimpanzee, and gibbon. They are tailless and are the largest of all primates, except for the gibbons, the smallest of the apes, considered the closest relatives of the human being.
Biped	An animal having only two feet.
Bipedal	Having two feet.
Bipedalism	The quality of being bipedal.
Bipolarized	From bipolar; having two poles.
Brainy	Intelligent, mentally acute.
Cerebralization	The process of becoming more cerebral or brainy.
Cetacea	Order of marine mammals with two suborders: *Mysticete*, including whalebone whales, and *Denticet*, including sperm whales and dolphins.
Chromosome	Any of the microscopic rod-shaped bodies found in the nucleus of the cell that carry the genes that convey hereditary characteristics. Chromosomes are constant for each species.
Diastema	The space between any two adjacent teeth.
DNA	An essential component of all living matter and a basic material in the chromosomes of the cell nucleus. DNA is responsible for the transmission of hereditary characteristics.
Embriology	A branch of biology dealing with the formation and development of embryos; the study of the way animals and plants develop during their earliest stages.
Evolution	The development of a species, organism or organ from its original state to the present state. Evolution is the theory, generally accepted, that all plants and animals developed from earlier forms by slight variations transmitted hereditarily in successive generations.
Gene	Any of the elements by which hereditary characteristics are transmitted and determined. They are found in the chromosomes.
Heredity	The transmission from parent to offspring of certain characteristics that produce the resemblance of the offspring to the parent.
Monkey	Any member of the *Primate* order, excluding lemurs, man, and the apes, or anthropomorphs.
Natural Selection	The process in nature by which individuals best fitted for the conditions of the environment in which their species live can survive, reproduce, and hence transmit these positive characteristics to their descendants.
Neoteny	The retention of infantile characteristics during adulthood.
Nidicolous	Adjective applied to any animal that needs a nest at birth, because of its little-developed condition.
Nidifugous	Adjective applied to any animal that does not need a nest at birth, because of its precocious development.
Ontogeny	The life cycle of a single organism; biological development of the individual; distinguished from *phylogeny*.

Pangenesis	Obsolete theory about the mechanics of heredity. Darwin accepted this theory because, in his time, the laws of heredity discovered by Mendel and the genetic science that followed his works were not yet known. Pangenesis , as understood by Darwin, considered that each cell of the body throws off very minute particles (gemmules) into the blood, which circulate freely and undergo division and are collected in the reproductive cells. Each part of the body is represented in the germ cells through these gemmules, considered the units of hereditary transmission. Darwin's intuition about heredity was oustanding because without knowing the laws of heredity and genetics, his conception of the mysterious process of hereditary transmission was very close.
Paradox	A statement apparently contradictory that can, in fact, be true.
Paradoxical	Having the nature of being or expressing a paradox.
Phylogeny	The racial history or evolutionary development of any plant or animal species.
Pinnipeda	Order of mammals; literally, "whit flippers on their feet;" include the seals, sea lions, and walruses.
Polarized	Having polarity; having the property of possessing two poles, positive and negative.
Proteins	Nitrogenous substances consisting of a complex union of aminoacids. Their are found in animal and vegetable matter. Apart from nitrogen, their major components are carbon, hydrogen, oxygen, and sulfur.
Puberty	The state of development in which an organism, animal, or person is able to reproduce.
Sirenia	Order of large vegetarian sea mammals, including the manatee and the dugong.
Taxonomy	The science of classification of living things (animals and plants). They are classified into : kingdom, phylum, class, order, family, genus and species.

Bibliography

Ainsworth, Mary D. 1964 "Patterns of Attachment Behavior by the Infant in Interaction with His Mother." *Merrill-Palmer Quarterly of Behavior and Development*, vol. 10, No. 1.

Ainsworth, Mary D. and Barbara A. Witting, 1968 "Attachment and Exploratory Behavior of One-Year Olds in a Strange Situation." *Determinants of Infant Behavior*. Vol. IV. New York: Willey.

Andrew, R. J. 1963 "Evolution of Facial Expression." *Science*, Nov. 22.

Andrew, R. J. 1962 "Evolution of Intelligence and Vocal Mimicking." *Science*, Aug. 24.

Andrew, R. J. *The Origin and Evolution of the Calls and Facial Expressions of the Primates.* Behavior , Vol. XX. 1963.

Andrew, R. J. 1965. "The Origins of Facial Expression." *Scientific American*, Oct.

Ardrey, R. 1976. "The Hunting Hypothesis." Collins, London: Collins.

Asimov, Isaac. 1965. "The Human Brain. Its Capacities and Functions." A Menton Book, New American Library.

Bailey, G. ed. 1983 "Hunter Gatherer Economy in Prehistory." Cambridge: Cambridge University Press.

Baroja, Julio Caro. "Razas, Pueblos y Linajes." Madrid: *Revista de Occidente*. Madrid.

Bartholomeu, George A. and Jodeph B. Birdsell. 1953. "Ecology and the Protohominids" *American Anthropologist*, Oct.

Bellugi, Ursula and Roger Brown, 1964. "The Acquisition of Language." *Monographs of the Society for Research in Child Development*. Vol.29, No. 1.

Binford, L. R. 1981. *Bones: Ancient Men and Modern Myths*. New York: Academic Press.

Bishop, W. W. ed. 1978. *Geological Background to Fossil Man*. Edimburg: Scottish Academic Press.

Boyce, A. J. ed. *Chromosome Variations in Human Evolution*. London: Taylor and Francis.

Brace, C. Loring. 1964. "A Consideration of Hominid Catastrophism." *Current Anthropology*. Feb.

Braidwood, Robert J. *1951. Prehistoric Men*. Chicago: Chicago Natural History Museum.

Brinton, Daniel C. 1946. *Raza Americana*. Buenos Aires: Editorial Nova.

Bronowsky, J. 1973. *The Ascent of Man*. London: BBC.

Brothwell, Don. 1963. "Where and When Did Man Become Wise?" *Discovery*, June.

Browman, D. L., ed 1978. *The Earliest Americans*. The Hague, Netherlands: The Hague, Mouton.

Bryan, A. L. (ed.) *Early Man in America From a Circumpacific Perspective*. Archaeological Researches International. Edmonton, 178.

Buettner-Janush, J. 1966. *The Origins of Man.* New York: Wiley.

Butzer, K. W. and G. L. I. Isaac. Eds. 1975. *After the Australopithecines.* Mouton, The Hague, The Netherlands: Mouton.

Campbell, Bernard. 1963. *Quantitative Taxonomy in Classification and Human Evolution.* Chicago: Aldine.

Childe, V. G. 1951. *Social Evolution.* Watts. London: Watts.

Chinese Academy of Sciences. 1980. *Atlas of Primitive Man in China.* Peking: Science Press.

Clark, W. E. LeGros. 1947. "The Importance of the Fossil Australopithecine in the Study of Human Evolution." *Science Progress*, July.

Clutton-Brock, J. 1981. *Domesticated Animals From Early Times.* London: Heineman and BMNH

Cole, Sonia. 1963. *The Prehistory of East Africa.* New York: MacMillan.

Coles, J. M. and E. S. Higgs. 1968. *The Archaeology of Early Man.* London: Faber and Faber.

Collins, Desmond. 1976. *The Human Revolution.* Oxford: Phaidon.

Coon, Carleston S., Stanley M. Garn and Joseph Birdsell B. 1950. *Races: A Study of the Problems of Race Formation in Man.* Springfield. Illinois: Charles C. Thomas.

Curtis, Garniss. 1961. "A Clock for the Ages: Potassium-Argon." *National Geographic.* Oct.

Curtis, Helena. 1978. *Biology.* New York: Worth Publishers, Inc. New York.

Darwin, Charles. 1874. *The Descent of Man.* London: John Murray.

Darwin, Charles. 1958. *The Origin of Species.* New York: The New American Library, Inc.

Darwin, Charles. 1977. *The Origin of Species and The Descent of Man.* New York: The Modern Library.

Day, M. 1977. *Guide to Fossil Man.* London: Cassell.

Dennell, R. 1983. *European Economic Prehistory: A New Approach.* London.

DeVore, Irven. 1963. *Maternal Behavior in Mammals.* New York: Willey.

DeVore, Irven. 1965. *Primate Behavior.* New York: Holt, Rinehart and Winston.

DeVore, Irven. 1968. *Man the Hunter.* Chicago: Aldine.

DeVore, Irven and S. L. Washburn. 1963. *African Ecology and Human Evolution.* Chicago: Aldine.

Dobzhansky, Theodosius. 1955. *Genética y el Origen de las Especies.* Madrid: Reviste de Occidente.

Eirmel, Sarel and Irven DeVore. 1965. *Primates* New York: Time-Life Books.

Eisenberg, John F. 1965. "The Social Organization of Mammals." *Handbuch der Zoologie*, Vol. 10, No.7.

Eldrege, Niles. 1985. *Times Frames. New York*: Simon and Schuster.

Ember, Carol and Melvin Ember. 1973. *Cultural Anthropology.* Englewood Cliffs: Prentice-Hall Inc.

Emperaire, J. 1955. *Les Nomades de la Mer.* Paris: Gallinard.

Fagen, Robert. 1981. *Animal Play Behavior.* New York: Oxford University Press.

Flannery, Kent. V. 1965. "The Ecology of Early Food Production in Mesopotamia." *Science*, March 12.

Fox, Robin. 1968. "The Evolution of Human Sexual Behavior." *New York Times Magazine*, March 24.

Fuentes, Jordi. 1965. *Prehispanic Textiles of Northern Chile.* Editorial A. Bello. Chile.

Fuentes, Jordi. 1960. *Grammar and Dictionary of the Easter Island Language.* Editorial A, Bello. Chile.

Fuentes, Jordi. 1985. *Man, the Paradoxical Primate*, Thousand Oaks: J.F.

Geschwind, Norman. 1964. "The Development of the Brain and the Evolution of Language." *Monograph Series of Languages and Linguistics.* No. 17. April.

Gili Gaya, Manuel. 1953. *Elementos de Fonética General.* Madrid: Editorial Gredos.

Goodman, Morris. 1962. "Immunology of the Primates and Primate Evolution." *Annals of the New York Academy of Sciences.* Dec. 28.

Gusdorf, Georges *La Palabra "La Parole."* Buenos Aires: Editorial. Galatea.

Hamburg, D. A. and E. R. McGown. 1979. eds. *The Great Apes.* Benjamin Menlo Park: Benjamin/ Cummings.

Harlan, J. R. J. M. J. de Wett and A. B. L. Stember. 1976. *Origins of African Plant Domesticatin.* The Hague, The Netherlands: Mouton.

Harlow, Harry F. and Margaret K. Harlow. 1962. "Social Deprivation in Monkeys." *Scientific American.* Nov.

Harlow, Harry F. and Robert R. Zimmermann. 1959. "Affectional Responses in the Infant Monkey." *Science.* Aug.

Harlow, Margaret K. and Harry F. Harlow "Affection in Primates." *Discovery.* Jan.

Haynes, B. F. 1982. "Human T Cell Antigen Expression by Primate T Cells." *Science*, Jan. Vol. 215, No. 4530.

Hays, H. R. 1964. *From Ape to Angel: an Informal History of Social Anthropology.* New York: Capricorn Books.

Hewes, Gordon W. 1964. "Hominid Bipedalism: Independent Evidence for the Food-Carrying Theory." *Science.* Oct. 16.

Hickman, Cleveland P. 1967. *Principios de Zoología.* Chile: Ediciones de la Universidad de Chile.

Higgs, E. S. 1972. *Papers in Economic Prehistory.* Cambridge: Cambridge University Press.

Hockett, Charles F. 1960. "The Origin of Speech." *Scientific American.* Sept.

Hockett, Charles F. 1968. "Comments on Current Trends in Linguistics." *Current Anthropology.* April-June.

Hockett, Charles F. 1964. and Robert Ascher. "The Human Revolution." *Current Anthropology.* June.

Hoebel, E. Adamson, Jesse Jennings D. and Elmere E. Smith. 1955. *Readings in Anthropology.* New York: McGraw Hill Book Co., Inc.

Holloway, R. L. 1967. "Human Aggression." *Natural History.* Dec.

House, Rafael Emilio. 1948. *Los Hijos del Sol.* Chile: Zig-Zag.

Howell, F. Clarl. 1971. *Early Man.* New York: Time-Life Books.

Howells, William. 1959. *Mankind in the Making,.* New York: Doubleday.

Huxley, J. 1974. *Evolution: The Modern Synthesis.* London: George Allen and Unwin.

Imbrie, J. and K. P. Imbrie. 1979. *Ice Ages: Solving the Mystery.* London: MacMillan.

Isaac, G. L. and E. R. McCown. 1976. *Human Origins.* Menlo Park: Benjamin.

Johanson, D. C. and M. A. Edey. 1981. *Lucy.* London: Granada.

Keeley, L. H. 1978. *Experimental Determination of Stone Tool Uses.* Chicago: University of Chicago Press.

Kellog, Winthrop N. 1968. "Communication and Language in the Home-Raised Chimpanzee." *Science.* Oct. 25.

Kenneth, D. Johnson, David L. Rayle, and Hale L. Wedberd. 1984. *Biology, An Introduction.* Menlo Park: The Benjamin/Cummings Publishing Co., Inc.

Keyser, Samuel J. 1964. "Our Manner of Speaking." *Technology Review.* Feb.

Knopf. Alfred A. 1984. *Ascent to Civilization, The Archaeology of Early Man.* New York: Alfred A. Knopf, Inc. new York.

Kortlandt. Adrian. 1962. "Chimpanzees in the Wild." *Scientific American.* May.

Kriekeberg, Walter. 1946. *Etnología Americana.* Mexico: Fondo de Cultura Económica.

Kroeber, A. L. 1953. *Anthropology Today.* Chicago: The University of Chicago Press.

LaBarre, Weston. 1956. *L'Animal Humain.* Paris: Payot.

Lancaster, Jane B. 1968. "Primate Communication Systems and The Emergence of Human Language in Primates." *Studies in Adaptation and Variability.* New York: Holt, Rinehart and Winston.

Lawick-Goodall, Jane van. 1967. "Mother-Offspring Relationships in Free-Ranging Chimpanzees." *Primate Ethology.* Chicago: Aldine.

Lawich-Goodall, Jane van. 1967. *My Friends, the Wild Chimpanzees.* Washington D. C.: National Geographic Society.

Lawick-Goodall, Jane van. 1963. *My life Among Wild Chimpanzees.* Washington D.C.: National Geographic Society. Aug.

Lawick-Goodall, Jane van. 1965. *New Discoveries among Africa's Chimpanzees.* Washington D.C.: National Geographic Society. Dec.

Leakey, L. S. B. 1960. *Adam's Ancestors.* New York: Harper & Row, New York.

Leakey, L. S. B. 1962. *Adventures in the Search for Man.* Washington : National Geographic Society. Jan.

Leakey, L. S. B. 1961. *Exploring 1,750,000 Years into Man's Past.* Washington: National Geographic Society. Oct.

Leakey, L. S. B. 1960. *Finding the World's Earliest Man.* Washington: National Geographic Society.

Leakey, L. S. B. 1954. ""Oldubai Gorge." *Scientific American.* Jan.

Leakey, Mary D. A. 1966. "A Review of the Oldubai Gorge, Tanzania." *Nature.* April 30

Leakey, Mary D. A. 1979. *Oldubai Gorge: My Search for Early Man.* London: Collins.

Leakey, Richard E. 1982. *Human Origins.* New York: Lodestar Books E. P. Dutton.

Leakey, Richard E. and Roger Lewin. 1977. *Origins.* New York: E. P. Dutton.

Lenneberg, Erick. 1967. *Biological Foundation of Language*. New York: Wiley.

Linton, Ralph. 1956. *Estudio del Hombre*. Mexico : Fondo de Cultuta Económica.

Lowie, Robert H. 1946. *Historia de La Etnología*. Mexico: Fondo de Cultura Económica.

Lowie Robert H. 1935. *Traite de Sociologie Primitive*. Paris: Payot.

MacDonald, Critchley. 1960. *The Evolution after Darwin*. Vol.2. Chicago: University of Chicago Press.

MacLean, Paul D. 1949. "New findings Relevant to the Evolution of Mind." *Quarterly Review of Biology*. March.

Marschack, A. 1972. *The Roots of Civilization*. London: Weidenfeld and Nicholson.

McNeill, Davis. 1966. "The Creation of Language." *Discovery*. July

Medina, Jose Toribio. 1952. "Los Aborígenes de Chile." Santiago: *Fondo Histórico y Bibliográfico José Toribio Medina*.

Moorey, P. R. S. 1979. "The Origin of Civilization." Oxford: Clarendon Press.

Napier, John. 1960. "Studies of the Hands of Living Primates." *Proceedings of the Zoological Society of London*. London: Zoologicall Society of London.

Napier, John, 1967. "The Antiquity of Human Walking." *Scientific American*. April.

Napier, John, 1962. "The Evolution of the Hand." *Scientific American*. Dec.

Nelson, Harry and Robert Jurmain. 1979. *Introduction to Physical Anthropology*. St. Paul: West Publishing Co.

Nordenskiol, Erland. 1956. Origen de las Civilizaciones Indígenas en la America del Sur. Buenos Aires: Editorial Bajel.

Oakley, Kenneth P. 1961. *Man the Tool Maker*. Chicago: University of Chicago Press.

Obermaier, Hugo, Antonio Garcia Bellido, y Luis Pericot. 1955. "El Hombre Prehistórico y los Origenes de la Humanidad." Madrid: *Revista de Occidente*.

Orr, William F. and Stephen c. Cappannari. 1964. "The Emergence of Language." *American Anthropologist. April.*

Osborn, Henry Fairchild. 1925. "Men of the Old Stone Age." New Yorl: Scribner.

Oage, J. W. 1941. "Les Derniers Peuples Primitifs." Paris: Payot.

Parr, A. E. 1967. "Urbanity and the Urban Scene." *Landscape*. Spring.

Pfeiffer, John E. 1962. "Man's First Revolution." *Horizon*. Sept.

Pfeiffer, John E. 1963. "The Apish Origins of Human Tension." *Harper's Magazine*. July.

Pfeiffer, John E. 1982. *The Creative Explosion*. New York: Harper & Row Publishers.

Pfeiffer, John E. 1978. *The Emergence of Man*. New York: Harper & Row.

Pfeiffer, John E. 1977. *The Emergence of Society*. New York: McGraw-Hill.

Pfeiffer, John E. 1955. *The Human Brain*. New York: Harper & Row.

Pfeiffer, John E. 1962. "Vision in Frogs." *Natural History*. Nov.

Pfeiffer, John E. 1965. "When Man First Stood Up." New York: *Time Magazine*. April.

Pfeiffer, John E. 1963. The Search for Early Man. New York: *American Heritage.*

Pilbeam, David R. 1967. "Man's Earliest Ancestors" *Science Journal.* Feb.

Premarck, Ann and David Premarck. 1982. *The Mind of an Ape.* New York: Norton.

Pritchard, E.E. Evans. 1957. *Antropología Social.* Buenos Aires: Editorial Nueva Visión.

Reader J. 1981. *Missing Links: The Hunt for Earliest Man.* London: Collins.

Reed, C. A., ed. 1977. *Origins of Agriculture.* The Hague, The Netherlands: Mouton. 1977

Renfrew, A. C. 1973. *Before Civilization.* London: Cape. London.

Reversz, G. 1950. *Origine et Prehistoire Du Langage.* Paris: Payot.

Reynolds, Vernon. 1964. "The Man of the Woods." *Natural History.* Jan.

Reynolds, Vernon. 1965. "Chimpanzees of the Budongo Forest." *Primate Beharvior.* New York: Holt, Rinehard and Winston.

Rivet, Paul. 1957. *Les Origines de L'Homme Americaine.* Paris: Gallinard.

Rowell, T. E. 1967. "Variability in the Social Organization of Primates." *Primate Ethology.* Chicago: Aldine.

Schneirla, T. C. 1966. "Instinct and Agrression." *Natural History.* Dec.

Schreider, Eugenio. 1950. *Los Tipos Humanos.* Mexico: Fondo de Cultura Económica.

Semenov, S. A. 1964. *Prehistoric Technology.* London: Cory, Adams and Mackay.

Service, Elman R. 1962. *Primitice Social Organization.* New York: Random House.

Sherratt, A., ed. 1980. *The Cambridge Encyclopedia of Archaeology.* Cambridge: Cambridge University Press.

Simons, Elwyn L. 1963. "Some Fallacies in the Study of Hominid Phylogeny." *Science.* Sept.

Simons, Elwyn, L. 1967. "The Earliest Apes." *Scientific American.* Dec.

Simons, Elwyn L. "The Early Relatives of Man." *Scientific American.* July.

Simpson, George Gaylord. 1949. *The Meaning of Evolution.* New Haven: Yale University Press.

Straus, William I. And A. J. E. Cave. 1957. "Pahology and the Posture of Neanderthal Man." *Quarterly Review of Biology.* Dec.

Tanner, J. M. 1968. "Earlier Maturation in Man." *Scientific American.* Jan.

Teilhard De Chardin, Pierre. 1957. *El Grupo Zoológico Humano.* Madrid: Taurus.

Tiger, Lionel. 1969. *Men in Groups.* New York: Random House.

Timbergen, N. 1964. "Aggression and Fear in the Normal Sexual Behavior of Some Animals." *Pathology and Treatment of Sexual Deviation.* New York: Oxford University Press.

Timbergen, N. 1968. "On War and Peace in Animals and Man." *Science.* June 28.

Tortora, Gerard J. 1983. *Principles of Human Anatomy.* New York: Harper & Row.

Ucko, P. J. and G. W. DIMBLEBY. 1969. *The Domestication and Exploitation of Plants and Animals.* London: Duckworth.

Van Valen, Leigh and Robert E. Sloan. 19965. "The Earliest Primates." *Science.* Nov.

Washburn, S. L. "Behavior and the Origin of Man." *Rockefeller University Review.* Jan-Feb.

Washburn, S. L. and Irven DeVore. 1961. "Social Behavior of Baboons and Early Man." *Social Life of Early Man.* Chicago: Aldine.

Washburn, S. L. and Irven DeVore. 1961. "The Social Life of Baboons." *Scientific American.* June.

Washburn, S. L. 1968. *The Study of Human Evolution.* Oregon: University of Oregon Press.

Washburn, Sherwood L. 1960. "Tools and Human Evolution." *Scientific American.* Sept.

Weinert, H. 1946. *L'Ascension Intellectuelle de L'Humanite.* Paris: Payot.

West, R. G. 1977. *Pleistocene Geology and Biology.* London: Longman.

Wood, B. A. 1978. *Human Evolution.* London: Chapman Hall.

Wood, Bernard. 1976. *The Evolution of Early Man.* London: Eurobook Limited.

Wormington, H. M. 1957. *Ancient Man in America.* 4th ed. Denver: Denver Museum of Natural History.

Wymer, J. 1982. *The Paleolithic Age.* London: Croom Helm.

Young, J. Z. 1974. *An Introduction to the Study of Man.* New York: Oxford University

Young, J. Z. 1965. "The Organization of a Memory System." *Proceedings of the Royal Society.* Vol. 163B. Nov. 23.

Young, J. Z., E. M. Jope and K. P. Oakley. eds. 1981. *The Emergence of Man.* London: Royal Society of London.

About the Author

Jordi Fuentes graduated from the University of Barcelona, Spain. He also graduated form the University of Chile, and is one of the founders of the Anthropological Studies Institute of Chile. He is noted as an anthropology specialist in the cultures of former Spanish Morocco.

Among his other achievements Mr. Fuentes has the following accomplishments to his credit:

Chief researcher for the National Historical Museum of Chile, and Curator of the Anthropological and Archaeological sections of this Museum.

Archaeologist in charge of the first textile research expedition in Northern Chile and Southern Peru cultures, and director of the first textile research and studies of the five most ancient cultures of Punta Pichalo, Pisagua.

Originator of research for the studies of Easter Island in the Anthropological Studies Institute of Chile.

Director of the Folkloric research team for the Andean-Pacific cultures in the National Historical Museum of Chile.

Director of the Historical studies on the Del Plata Vice royalty and Chilean Congress Library Comptroller for the Chilean Encyclopedia (Enciclopedia Chilena), and author of several books on his specialties.

Previous Publications
By
Jordi Fuentes

GRAMMAR AND DICTIONARY OF EASTER ISLAND LANGUAGE
Editorial Andres Bello, Santiago de Chile, Chile, 1960

PREHISPANIC TEXTILES OF NORTHERN CHILE
Editorial Andres Bello, Santiago de Chile, Chile, 1966

HISTORICAL DICTIONARY OF CHILE
Jordi Fuentes & Lia Cortes
Editorial Del Pacifico, Santiago de Chile, Chile, 1966

POLITICAL DICTIONARY OF CHILE
Jordi Fuentes & Lia Cortes
Editorial Orbe, Santiago de Chile, Chile, 1967

HISTORICAL DICTIONARY OF THE UNITED STATES
Alfa Publications, Los Angeles, California, U.S.A. 1978

MAN, THE PARADOXICAL PRIMATE
A Solution to Human Bipedalism
Jordi Fuentes, Thousand Oaks California, U.S.A. 1985

QUESTIONS AND ANSWERS ABOUT CACTI AND SUCCULENTS
Jordi Fuentes, Thousand Oaks California, U.S.A. 1994

FOLKLORIC DICTIONARY OF CHILE
Legends, Believes, uses, and traditions
Jordi Fuentes, Thousand Oaks California, U.S.A. 1994

SEXUALITY AND COMMUNISM
Biological Causes for the Downfall of Communism

www.ingramcontent.com/pod-product-compliance
Ingram Content Group UK Ltd.
Pitfield, Milton Keynes, MK11 3LW, UK
UKHW061831190726
13855UKWH00005B/1745

9 781587 214868